Stefan Baerisch

Domain-Specific Model-Driven Testing

VIEWEG+TEUBNER RESEARCH

Software Engineering Research

Herausgeber/Editor:

Prof. Dr. Wilhelm Hasselbring

Im Software Engineering wird traditionell ein Fokus auf den Prozess der Konstruktion von Softwaresystemen gelegt. Der Betrieb von Systemen, die kontinuierlich Dienste mit einer geforderten Qualität bieten müssen, stellt eine ebenso große Herausforderung dar. Ziel der Reihe Software Engineering Research ist es, innovative Techniken und Methoden für die Entwicklung und den Betrieb von nachhaltigen Softwaresystemen vorzustellen.

Traditionally, software engineering focuses on the process of constructing and evolving software systems. The operation of systems that are expected to continuously provide services with required quality properties is another great challenge. It is the goal of the Series Software Engineering Research to present innovative techniques and methods for engineering and operating sustainable software systems.

Stefan Baerisch

Domain-Specific Model-Driven Testing

VIEWEG+TEUBNER RESEARCH

Bibliographic information published by the Deutsche Nationalbibliothek
The Deutsche Nationalbibliothek lists this publication in the Deutsche Nationalbibliografie; detailed bibliographic data are available in the Internet at http://dnb.d-nb.de.

Dissertation Universität Kiel, 2009

1st Edition 2010

Editorial Office: Ute Wrasmann | Anita Wilke

Vieweg+Teubner is part of the specialist publishing group Springer Science+Business Media.
www.viewegteubner.de

Cover design: KünkelLopka Medienentwicklung, Heidelberg
Printed on acid-free paper
Printed in Germany

ISBN 978-3-8348-0931-5

Foreword

Meeting the requirements on software systems is a great challenge in software engineering. To meet the *real* user requirements, domain experts need to participate in the software engineering process. The research work presented here addresses the integration of domain experts into the testing activities, particularly with respect to acceptance tests.

Stefan Baerisch presents with MTCC (Model-Driven Test Case Construction) an innovative approach to formally specifying automated acceptance tests in the application domain of digital libraries. Digital libraries may be grouped into a system family of software systems sharing some commonalities as well as some variability. With MTCC, test cases are generated with model-driven techniques within the context of the digital-libraries system family. With MTCC, domain experts are involved into the testing process without requiring programming capabilities from these domain experts.

In software engineering, it is not sufficient to propose new engineering techniques. It is required to empirically evaluate the new techniques. The work presented in this book does not only present the new MTCC approach, it also reports on a significant empirical evaluation of MTCC by domain experts at the GESIS institute. Particularly, the detailed documentation of the validation goals and the according experiment design with a candid discussion of the threads to validity may serve as an example for other software engineering research projects—which is why this book is a valuable addition to a researcher's library.

Wilhelm Hasselbring

Acknowledgements

Many people deserve my gratitude for the support and encouragement they gave me during my work on this work.

My advisor, Professor Dr. Wilhelm Hasselbring, helped me to shape and refine the ideas underlying this work and supported me in developing the means to express, implement, and validate these ideas in this dissertation. Thank you.

Professor Dr. Jürgen Krause gave me the opportunity to combine my studies and work on this work with the always interesting work for the GESIS. I thank him for his support and his belief in this work as well as for our interesting conversations.

I thank all my colleagues at the GESIS, both past and present. Max Stempfhuber gave me the freedom to combine this dissertation with my other responsibilities and pointed out areas of improvement. Patrick Lay and Holger Heusser gave me important feedback on my thoughts.

Thanks also to all domain experts. I am grateful for the time and effort contributed by the members of the evaluation group, Vivien Petras, Maria Zens, Simone Weber, and Jan-Hendrik Schulz.

Rielies Neitzke proved that a librarian can be found at the heart of all research. Harald Deuer and Maria Zens always had time for some words. My deepest thanks to Vivien Petras for our discussions as well as for her support and encouragement.

It is always worthwhile to present ones ideas to a fresh audience. My thanks to the ESEC/FSE 2007 Doctoral Symposium for their time and insights, in particular for emphasizing the importance of a sound evaluation.

The Software Engineering Group at the Carl von Ossietzky University of Oldenburg gave me the opportunity to present and discuss many of the ideas that are presented in this work. I thank you both for this and for the friendly atmosphere I could always count on.

My parents made this dissertation possible in more than one way. Elke and Wolfgang, you have my sincerest thanks and gratitude for all you have done in the past 31 years.

Stefan Baerisch

Abstract

This work presents MTCC (Model-Driven Test Case Construction), an approach to the construction of acceptance tests by domain experts for testing system families based on feature models. MTCC is applied to the application domain of Digital Libraries.

The creation and maintenance of high quality systems that fulfill both formal requirements and meet the needs of users is one of the primary goals of software engineering. A prerequisite for the quality is the absence of faults. Software engineering has defined a number of techniques for avoiding faults, identifying them or fixing them. Testing identifies faults by exercising an implementation artifact and comparing its actual and expected behavior. MTCC is an approach to automate acceptance tests for the members of system families.

The basic hypothesis of this work is that the involvement of domain experts in the testing process for members of system families is possible on the basis of feature models and that such a testing approach has a positive influence on the efficiency and effectiveness of testing. Application quality benefits from the involvement of domain experts because tests specified by domain experts reflect their needs and requirements and, therefore, can serve as an executable specification.

One prerequisite for the inclusion of domain experts is tooling that supports the specification of automated tests without formal modeling or programming skills. In MTCC, models of automated acceptance tests are constructed with a graphical editor based on models that represent the test-relevant functionality of a system under test as feature models and finite state machines. Feature models for individual testable systems are derived from domain-level systems for the system family. The use of feature models by the test reuse system of MTCC facilitates the systematic reuse of test models for the members of system families. MTCC is a model-driven test automation approach that aims at increasing the efficiency of test execution by automation while keeping independence from the implementation of the testee or the test harness in use. Since tests in MTCC are abstract models that represent the intent of the test independently from implementation specifics, MTCC employs a template-based code generation approach to generate executable test cases.

In order to validate the approach, MTCC is applied to the Digital Library application domain. Digital Libraries are Information Retrieval systems that aim at providing the scientific community with relevant information. A MTCC prototype is designed and realized for a system family of three Digital Libraries and an Information Retrieval system. The capability of representing tests relevant for the application domain, for reusing these tests for multiple systems and for generating executable tests from the abstract test models are validated. An assessment of the understandability by domain experts and of the usability of the editor is conducted. The feasibility and practicality are shown by a validation involving domain experts for a system family of Digital Libraries.

Contents

1 Testing of Software

Testing is defined as the execution of a system with the purpose of finding faults in the system. It encompasses a wide range of different approaches with different goals and motivations. In this chapter, we discuss both the role of testing for software quality and the role of MTCC in the field of testing. We start with an examination of the role of testing in Section 1.1.

In order to be more efficient and effective, often, various parts of the testing process are automated. The extent, to which testing is automated, varies with different approaches, but the execution of tests implemented as test scripts and the recording of test results form the core of test automation. Section 1.1.3 examined the goals of test automation and the challenges that must be met to establish an automated testing regime.

Acceptance tests involve domain experts in system testing. Section 1.2 motivates the involvement of domain experts into the testing process and discusses acceptance tests and their automation.

The quality of a test is defined by its ability to find faults. Tests must therefore be based on the requirements as defined by the domain experts for a system and must be able the assess a testee based on this requirements. Section 1.3 discusses the meaning of quality for tests and test software and applies the quality attributes of software engineering to test software.

1.1 Testing and Software Quality

Testing is a method of quality assurance that verifies the behavior of a system against a set of requirements that are expressed as tests. The notion of quality and its relation to the requirements of a system, the specifications that represent these requirements and the needs of the users of a system are central to testing.

If quality is defined as the capability of a system to fulfill the requirements of its users or customers [JG88], the question must be considered how these requirements can be captured and formalized as tests. One source of requirements are domain experts. Domain experts are customers or users of a system who are familiar with the functions that the systems needs to fulfill and who are often experienced in the use of similar systems.

The direct and continuous involvement of domain expert in the software development process as a methodology belongs to the group of agile approaches to software development [BA04] but is considered as part of other approaches to software development as well [Eva04].

One method to make the knowledge of domain experts available for the software development process is the definition of acceptance tests. In acceptance tests, domain experts either test a system themselves or specify the functionality to be tested and the conditions for the success of a test in a format that is either suitable for execution by a human tester or for execution by an automated testing system.

One purpose of test automation is to allow the execution of tests without human involvement. Automation is desirable both for economic reasons and to allow for greater effectiveness in testing.

1.1.1 Software Quality

Juran [JG88] defines quality as *fitness for use*. This definition raises a number of questions: fitness for what use? What constitutes fitness? How can it be determined, if and to what degree a system is fit to be used? In order to evaluate the fitness of a system, some sort of specification is needed. A specification is a document that describes the requirements for a system. The quality of a system can also be defined as the degree to which its implementation complies with its specification. The specification must therefore reflect the requirements for a system.

Functional as well as non-functional quality attributes are relevant to software systems. Software engineering defines a number of methods that serve the purpose of creating a high-quality products or evaluating its quality.

1.1.1.1 Different Notions of Quality

Two questions have to be addressed in the discussion of quality assurance. At first, quality has to be defined. Based on this definition, methods and tools have to be identified that allow for the evaluation of quality and thereby provide a prerequisite to increase quality.

Two basic notions of quality exist [RB94]: The view that quality is the adherence to a specification or that quality is the degree to which a product or Service is useful to its users or customers. The definition of quality given by Crosby [Cro79] is exemplary for the former view on quality while Edwards [Edw68] and Juran [JG88] are proponents of the latter. We call these different notions of quality specification-centric and user-centric.

Applied to the field of software engineering, the specification-centric and the user-centric notions of quality are reconciled [Kan02] when the specification of a system represents the needs of its users completely and correctly. The quality of a specification is instrumental for the quality of the system built based on this specification. The quality of a specification is determined by the degree to which it reflects the users requirements, needs, and to a certain degree wishes. In order to build high-quality systems, both high-quality specifications and ways to compare a specification and the implemented systems are needed.

Validation is the activity to ensure that *the right things are done*, that is, a specification represents the requirements of a system. Verification aims to ensure that *things are done right* and that a system reflects all points of the specification. Testing is foremost a verification activity. It systematically compares a specification with a system; tests are examples for the expected behavior of the system defined by the specification.

We argue that tests can also serve as a partial specification, a *specification by example* [Fow06], in particular in situations where the actual specifications are incomplete or missing. In such situations, tests cannot be derived from specifications but must be derived from the requirements of a system and must be specified by users or customers. It follows that such acceptance testing by users who are experts for the subject domain is a validation activity.

1.1.1.2 Attributes of Software Quality

The concept of software quality has multiple dimensions or attributes. The ISO standard 9126-1 [fS01] defines the quality of software along six attributes: Functionality, Reliability, Usability, Efficiency, Maintainability, and Portability.

We distinguish between functional and non-functional quality as well as between user-observable quality and not user-observable quality [McC04]:

- Functional quality is the capability of a system to correctly fulfill a specific task. Non-functional quality attributes describe general conditions that the system must meet, for instance limits to the time and resources used or the reliability of a system. The final five of the six ISO quality attributes given above describe the non-functional qualities of a system.

- The user-observable quality of a system includes those attributes that effect a user working with the system. User-observable quality subsumes both functional and non-functional quality attributes but excludes quality attributes such as maintainability or server-side resource usage that are not visible to a user.

The notion of non-user-observable quality appears contradictory based on our definition that quality is, either direct or indirect, the degree to which a system meets the requirements of its users. It is indeed the case that quality attributes like maintainability are not visible to a user and are yet crucial for the sustained quality of a system. The existence of non-user-observable quality attributes is one reason why acceptance tests are only a partial specification. We argue that this limitation does not challenge the suitability of tests as a specification for the user-observable quality attributes

1.1.1.3 Methods of Quality Assurance

There is no method or tool in software engineering that can serve as a silver bullet [Bro87] and guarantee high quality throughout the whole software development process and for all products. As a consequence, software engineering employs a bundle of techniques that support the construction of high-quality artifacts or evaluate the quality of existing products and processes.

Among the measures that directly influence the quality of a product under development are methodologies and software process models, for instance the V-Model [BR05], the Rational Unified Process[Kru03, Lar04] or agile methods [BA04] and the utilization of established best practices like design patterns [BMR$^+$96, GHJV95] as well as the use of tools that support the software development process like version control systems [PCSF04].

While the measures above address the creation and increase of quality in the product development process, complementary means are necessary to verify the quality of a product and identify quality deficits. Testing is one means for this purpose.

1.1.2 Definition of Testing

A basic question about the role of testing is whether it serves to ascertain the quality of a product or identify faults. Fault-oriented testing and conformance-oriented testing are contradictory views on the role of testing and give rise to the question exactly how testing benefits the quality of a software system and what characterizes good testing.

1.1.2.1 Conformance and Fault-Oriented Testing

Conformance-oriented testing is based on the assumption that the purpose of testing lies in proving or at least demonstrating the quality of product. Approaches that advance a conformance-oriented view of testing usually place testing towards the end of the development of a system or subsystem, for example, at the end of a naive implementation for the waterfall [Roy87] model of software development.

The focus of fault-oriented testing, in contrast, lies on the identification of faults in the system. Since testing can never prove the absence of faults [Dij69], the fault-oriented view on testing only regards those tests as useful that find faults [MSBT04].

The current view on testing is a compromise of the conformance-oriented and the fault-oriented view. While the notion that correctness cannot be proven by testing is not challenged, the usefulness of tests that do not find faults is also accepted [Che05a]. While, for instance, regression tests are not expected to find faults in a system, they provide an assurance against faults that were introduced in existing systems. While tests cannot prove the absence of faults, high test coverage of a code base both increases the probability to find faults and allows for some degree of confidence about the likely faultiness of a system or subsystem.

Beside the fault-oriented and conformance-oriented views on testing, yet another view is held by the Test Driven Development [Bec02, JS05] approach. In TDD, tests serve to clarify the requirements that exist for a software and ensure that the intended behavior of an implementation artifact matches its actual behavior. The goal of testing in TDD is not to find faults but to guide the development of a system. *"One of the ironies of TDD is that it isn't a testing technique. It's an analysis technique, a design technique, really a technique structuring all the activities of development"* [Bec02].

In the context of MTCC we advance the view that automated acceptance tests both serve to describe the intended behavior of a system and to find faults in a system. MTCC tests also serve as regression tests and thus, under the precondition that test coverage is sufficiently high, provide some confidence that a system is free of defects.

MTCC shares one aspect with TDD in the fact that tests serve as a means to express requirements for a system, it is however no test-first approach since a testable system must exist prior to acceptance testing. Also, acceptance testing with different roles does not support the *red-green-refactor* cycle that Beck describes as the mantra of TDD. We regard testing with MTCC as a supplement to fault-oriented testing that raises quality both by supporting acceptance tests and by facilitating to identify conflicts between the requirements of domain experts and an implementation.

1.1.2.2 Quality of Tests

The ability to identify faults in a system and to provide means to assess the likeliness of faults with some confidence do not in themselves lead to criteria to judge the quality of a test. In addition to the effectiveness of a test, its ability to find faults, non-functional criteria like efficiency and maintainability also contribute to the quality of a test.

Graham [GF00] introduce four attributes of quality for automated tests that can be applied to tests in general: A test is *effective*(1) when it is likely to find faults in a system, it is *exemplary*(2) when it does not only cover highly specific functionality of a system but tests multiple aspects of a system at once.

The resources needed to design, implement, execute and debug the test describe how *economical*(3) a test is. A test is *evolvable*(4) when it can be maintained and adapted to changes in the testee with high costs.

Figure 1.1 illustrates how the four quality attributes defined by Graham and Fewster apply to manual and automated testing. The attributes help to illustrate that testing is an activity that has economic as well as purely functional aspects. The economic side of testing is one of the main motivations behind the automation of tests.

1.1.3 Automated Testing

One goal of MTCC is the automation of acceptance tests. In the following we discuss in detail why we argue that such automation is beneficial and which activities of testing are suitable for automation. We regard test automation as advantageous with respect to both the effectiveness and the efficiency of testing, in particular in the context of regression tests. We argue that the subject knowledge and understanding for the requirements of a system that domain experts can provide is better suited for the testing of digital libraries than the complete automation of the test design phase. We therefore provide an editor to support the construction of tests.

1.1.3.1 Reasons for Test Automation

Since we regard testing as an activity that allows the assessment of the quality of a system during the whole software life cycle, it follows that testing and in particular test execution are activities that have to be supported and thus consume resources over the whole length of a product development and maintenance cycle. Test execution can be very frequent, continuous integration [DMG07] approaches, for example, aim to execute each test at least daily.

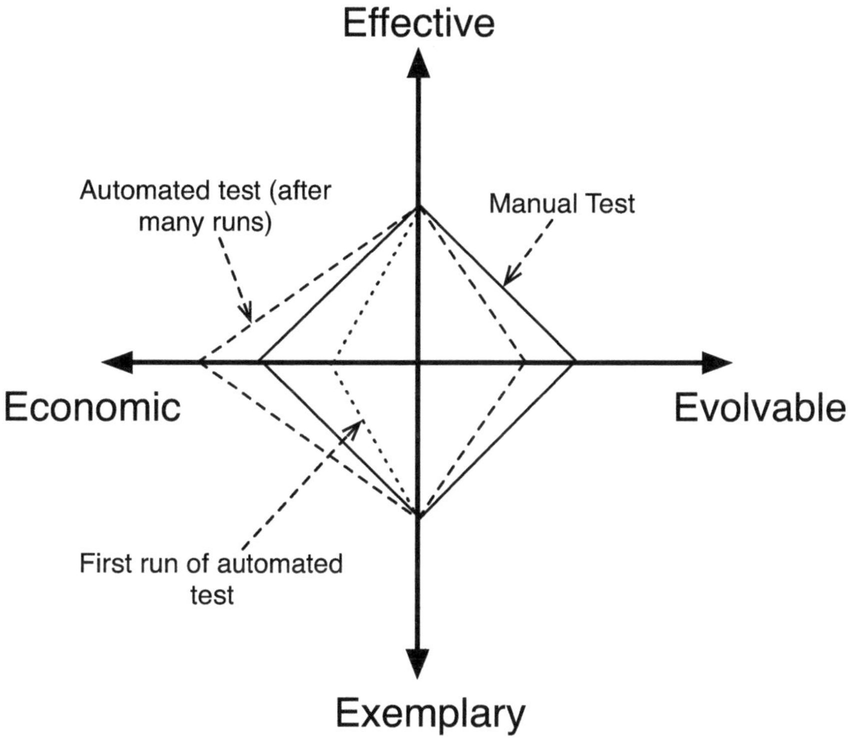

Figure 1.1: Attributes of good tests [GF00]

Test execution with such frequency cannot be achieved manually in practice for a non-trivial number of tests, the automatization of test execution therefore is imperative. Test automation is advantageous for a number of reasons even when tests are less frequently executed [GF00]:

- Automated tests have economic advantages compared to manually executed tests. Manual testing requires human involvement and is thus costly while the execution of test cases will generally consume only minimal resources. For this reasons, the costs for the implementation of automatic testing will usually be lower than those for manual testing when tests are repeatedly executed.

- The execution of automated tests does not conflict with the day-to-day tasks of software engineers or domain experts. This focuses on the fact that the resources for manual test execution need to be available in all organizations and all phases of the software development process. If, for instance, manual test execution is done by software developers, test execution and software development will both compete for the time of this person. While the implementation of test automation software will also require resources from a software developer, this is a one-time effort.

- Automated tests are potentially more effective than manual tests. Automated tests support the execution of more tests more often. Such tests are also more reproducible than manual tests since no variances are introduced into the test execution process by human testers.

The above claim that a higher frequency of test execution, a greater number of tests and better reproducibility result in an overall increase in test effectiveness requires further consideration. We argue that such an increase occurs because of the following points:

- The number of tests and the frequency with which they can be executed by automated test software is not achievable by manual test execution. The number of automated tests that can be executed in a given amount of time is limited only by the available computation resources. While the number of executed tests alone has little influence on the ability of a test regime to find faults, the ability to frequently execute all tests for a given system greatly increases the effectiveness of regression testing.

- Automated tests are reproducible [Bin99]. Using test automation it is possible to exactly specify and repeat the inputs send to test the testee during test execution. This allows for the exact manipulation of the state of the testee and supports the execution of repeatable tests.

- Automated tests support the verification of quality attributes that can not be tested by manual tests in practice. Load tests that require a large number of requests to be send to the testee within a short period of time can not be realized without test automation.

We conclude that test automation has high potential to increase the effectiveness and efficiency of testing. Binder goes so far as to call automation the central aspect of testing:*"(Testing) is the design and implementation of a special kind of software system: one that exercises another software system with the intent of finding bugs"* [Bin99]. We advance the view that test automation primarily benefits the efficiency and effectiveness of test execution but that explorative testing [KBP01] is necessary in addition in order to identify new tests for automation. It is therefore one purpose of MTCC to ease the automation of test cases identified by explorative testing by providing an environment for test modeling.

1.1.3.2 Phases of Test Automation

Our discussion of test automation so far was limited on the automation of test execution. This is justified by the fact that test execution is the phase of testing that is most often automated. Phases like test design and test implementation are far less frequently automated in practice.

Fewster and Graham [GF00] partitions the automated testing process in five phases: The identification of test scenarios, the design of tests, the implementation of test automation, the execution of tests and the comparison of actual and expected results. Figure 1.2 illustrates these activities and their potential for automation:

- The purpose of the Identify phase is the identification of functionality that can be tested.
- The goal of the Design phase is to produce concrete tests for the functionality identified during the Identify phase. Input values and expected results are specified for each test.
- During the Build or implementation phase, the tests designed in the previous phase are implemented in a notation suitable for automatic execution, for instance in a scripting language.
- The Execute phase sees the execution of the previously implemented tests on the system under test.
- The purpose of the Check phase is the evaluation of the success of the test based on a comparison of the expected and the actual behavior of the SUT.

As Figure 1.2 illustrates, the various phases of testing vary in their suitability for test automation. The automation of the test execution phase and the comparison of actual and expected results can generally be automated without significant problems, the automation of this phases also results in large practical benefits since both phases are frequently executed in the testing process.

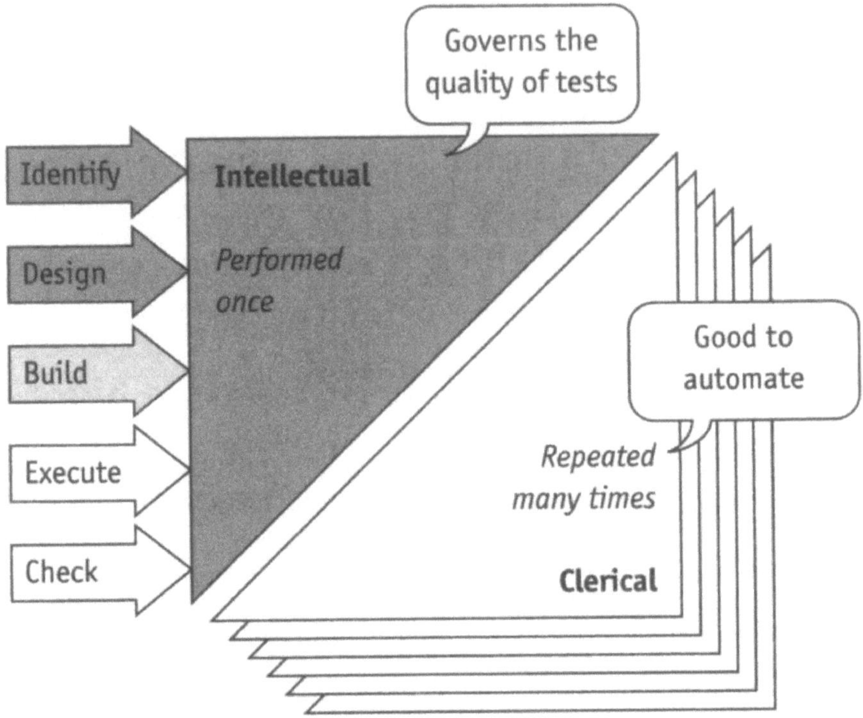

Figure 1.2: The five phases of testing [GF00]

MTCC automates the test execution and the Comparison phase. The build phase is also automated once the Test Adapter and the necessary templates are implemented. The design phase is not automated but is supported by the editor.

1.1.3.3 Automatic and Automated Testing

In order to distinguish MTCC from other approaches for test automation, in particular from model-driven testing (discussed in Section 2.3), we differentiate between automated testing and automatic testing.

The purpose of automated testing is the automation of test execution and comparison and in some cases the build or test implementation phase. Automatic testing also aims at the automation of the test design phase. We consider automated testing not as a partial implementation of automatic testing but as a separate field. Since test design in automated testing is done by humans, the support of intellectual test design, not its automation, is the focus of the effort. Specifically, automatic and automated testing differ in the following points:

- Automatic testing requires a formal model of the testee [BMM04] that represents all states of the testee as well as events that can influence this state, automatic testing is therefore always also model-driven testing.

  Automated testing in contrast is based on the knowledge of domain experts; the approach does not require models at the level of detail necessary for automatic testing. We argue that automated testing is better suited for situations where testing is mostly based on the requirements of users, not formal specifications.

- Automated and automatic testing do not differ in test execution. Both approaches use an automated test runner to execute a test case without human involvement.

- A relevant difference between automated and automatic testing is the approach used to determine the success of a test. In automated testing, the conditions for the success of a test are defined by a human during test design, automatic testing utilizes an oracle [BY01] for the decision if a test was successful or not. Depending on the testee and the conditions to be tested, the complexity of an oracle can vary widely. An oracle that verifies that the testee did not crash during test execution is far less complex than an oracle that verifies complex business processes [MBN03]. If an alternative implementation of the system under test exists, it can serve as an oracle in the context of differential testing [McK98, ES07]. The fact that automated testing can rely on human intellect to describe the success criteria of tests allows for the testing in situations where the automatic determination of test success is not possible, for instance in the context of Digital Libraries.

We argue that automatic and automated testing supplement each other. We further argue that the inclusion of humans in the testing process leads to advantages that cannot be achieved by automatic testing: While automatic testing is dependent on the correctness and completeness of the specification upon which the models used for testing and the oracle are built, automated testing allows for the discovery of new requirements in the testing process.

1.2 Acceptance Testing

Given that the intent of testing as a verification activity is to ensure that the implementation of a system conforms to its specification, and that a specification is meant to represent the requirements of the users of a system, it appears to be beneficial to the quality of software to directly involve domain experts in testing and thus ensure that the tests executed on the system represent the real requirements to the system.

Domain experts are assumed to have detailed knowledge of the subject matter and the functional requirements for a system, but not to have any skills in formal modeling, software development, or software testing [AB04, And04a, MC05b].

1.2.1 Testing by Domain Experts

Tests that directly involve domain experts in system testing are called acceptance tests [MC05b]. Domain experts can participate in acceptance testing in a number of different ways, They can manually design and execute tests, design tests that are automated by software developers or use specialized tools for acceptance testing such as FIT or Selenium. One purpose of the MTCC approach presented in this work is to provide a tool that allows the specification of tests by domain experts in the context of system families.

We consider acceptance testing an activity in parallel to the software development process that starts as soon as a first testable prototype of the system is available. The advantage of such an approach over acceptance testing only at the end of the software development process is that the tests can facilitate discussions over the quality of the system and help discover previously unknown requirements.

Compared to the evaluation of the degree to which a system meets its requirements purely based on a specification, testing has a number of advantages: Since testing is always specific for a concrete, existing system and since tests need to be detailed enough for automated execution, the preparation of tests supports the discovery of omissions or errors in the requirements better than a informal specification.

The inclusion of domain experts or customers in the software development process is explicitly demanded in agile methodologies like Extreme Programming [BA04], often in a testing role. The inclusion of customers in the testing is generally considered desirable [AOS04].

1.2.2 Types of Involvement

Since in most cases domain experts will not be professional testers nor software developers, domain experts cannot automate the execution of acceptance testing themselves by expressing their tests as test cases. They can instead participate in one of the following ways:

- Acceptance tests can be specified by domain experts in a informal way and are then implemented by software developers. The approach has the disadvantage that domain experts are not directly involved in the testing process. In order to ensure that tests represent the intentions of the domain experts, the test specification must be detailed enough to prevent misunderstandings. Such a specification must be as formal as the test it represents, it merely omits implementations details. As a consequence, test and specification are redundant.

- Acceptance tests are specified using a detailed but informal test plan, the execution of tests is done by professional testers. While there is no redundancy between a test implementation and the test specification — there is no implementation — this approach suffers from the disadvantages of manual test execution.

- Domain experts can test the SUT themselves, either by explorative testing or following a simple test plan. The main disadvantage of this approach is the increased work load on the domain experts, especially considering the fact that test will be executed frequently.

- Domain experts construct tests suitable for automatic execution themselves, using a tool usable without further training. This approach combines the advantages of automated testing and the involvement of domain experts in the testing process. This approach is followed by MTCC.

1.3 Quality of Test Automation

So far we have introduced our notion of quality for software and discussed criteria to evaluate the quality of tests, but we have not yet applied our criteria for quality to testing software itself.

In consideration of the fact that test automation software is software and thus can be judged in terms of its functional and non-functional qualities, we now discuss what determines the quality of automated acceptance tests. From the result of this discussion we derive the requirement of the MTCC approach.

1.3.1 Quality of Requirements and Quality of Tests

As always in software engineering, the quality that can be reached in any phase of the development of a system for testing is limited by the results of the proceeding phase. Without thorough requirements engineering, it is not possible to design the right system and without design, the results of the implementation are more than in question. This applies for incremental process models as well as the traditional V-Model.

The consequence for test software is, that while domain experts must be present at all phases of the development of test software, their presence is particularly important in the early phases of the development of test software when the functionality of a testee that can be tested and the basic capabilities of the test software are determined. The development of test software needs the cooperation of experts on the subject matter to be tested and on the implementation of the testee as described, for instance, in the concept of Domain Driven Design [Eva04].

In the context of MTCC, this means that domain experts are not only involved in the construction of tests but also in the analysis of the application domain/system family and the individual systems within the system family.

1.3.2 Software Quality Attributes Applied to Test Software

The six quality attributes defined by the ISO standard 9126 can be applied to automated tests as well as to regular software. Non-functional quality attributes, such as the maintainability and usability, have a direct impact on the fitness of test software to verify the quality of a testee.

In the context of software product-lines, it is of special importance that test software can be adapted to the specifics of different testees. In the particular case of MTCC, the usability of those aspects of the test software that are intended for use by domain experts are highly relevant to the ability of the system to test the members of the product family.

While the quality attributes for automated test systems are the same as those for regular software, special priorities apply. We argue that a system that aims to involve domain experts in the testing process must meet the following requirements:

- The functionality of the system must include the capability to cover all fault scenarios for the system family under consideration. The test automation system must include functionality to take a representation of a test in some formal notation, execute the test on the testee and decide on the success or failure of the test.

- In terms of usability, a system for the specification of automated acceptance tests for domain experts must consider a number of different aspects, the most important of which is the language or the editor that domain experts use for the specification of tests.

- The efficiency of an automated test system is defined by its ability to execute a large volume of tests with limited resources. The efficiency of the execution of an automated test depends on the resources necessary for the execution of the testee since mocking [Mes07] is not possible in a system test context.

- Since test software needs to be adapted to changes of the testee that affect the testable interface of the testee, maintainability and adaptability are important attributes for such systems. Maintainability and adaptability are of special importance in a system family context when a sustainable test process must be established not only for a single system but for multiple different implementations. Beside the ability to follow changes in the testee, the test automation software must also be maintainable in the traditional sense, for example in order to support refactoring [DMBK01, Mes07].

- An important quality attribute for a test software in a system family context is the ability to reuse tests. Reuse means that a test that was specified for one system can be transfered to and executed on another system if this system has the necessary functionality for the test to be executed.

We argue that all functional and non-function quality attributes are of high relevance for test automation software and that non-functional quality attributes have a significant impact on the overall effectiveness of such a system. For instance, we consider it highly unlikely that the specification of acceptance tests is practical if the user interface used for the specification is not usable. MTCC therefore aims to provide an editor that facilitates the construction of test cases.

1.4 Chapter Summary

Testing is one of the methods of quality assurance for software. It differs from other methods, for example Static Analysis, in the fact that it executes the code of the SUT and compares the encountered behavior of the SUT with its expected behavior. The quality of a test is defined by a number of factors. A prerequisite for a high-quality test is that it represents the requirements for the SUT and can be used to verify these requirements.

A test must also be efficient — it must be possible to execute a test with minimal resources. One way to minimize the resources necessary for test execution is test automation. Acceptance Tests involve domain experts in the testing process. Since domain experts are not professional software developers or testers, special tool support is necessary to support domain experts in the specification of automated acceptance tests.

As software systems, automated tests are subject to all software quality attributes defined in the ISO 9126. In addition to these quality attributes, some specific requirements apply to a test automation system in a system family background. Such a system must support tests that are reusable for multiple systems and can be adapted to changes in the SUT.

2 Use of Models in Software Engineering

In this chapter we discuss the use of models in software engineering. We examine the relation of MTCC to the fields of model-driven software development and model-driven testing.

Based on the definition of Stachowiak [Sta73] of models as abstractions with a purpose and a representation, the use of models is a well established practice in software engineering as well as in computer science as a whole. Models are not only used in approaches that are explicitly model-driven but are employed in a number of different forms and for a multitude of tasks. We consider an approach model-driven if models are the primary artifacts in a process from which the different implementation artifacts are generated. Models are used to abstract data and control flow in programming languages [Sco05]. The design of a suitable domain model [Fow03] is a central part in the development of complex software systems and the elicitation of a suitable fault model [Bin99] is a central part of the testing process. Section 2.1 discusses the definition of models we employ for MTCC and examines the notions of meta models and model transformations.

Current model-driven processes and technologies are characterized by the utilization of formal models on a level of abstraction above the specific implementation platform and by the use of model transformation, generally for the purpose of code generation, in order to bridge the implementation gap [FR07] between an abstract model and a specific platform.

A number of different approaches for model-driven development exist that influence the MTCC approach to different degrees. MTCC is closest to the generative programming approach regarding the used models and the high degree to which the details of the implementation domain are abstracted. We introduce these approaches in Section 2.2 and compare them with MTCC.

One concrete use of models is model-driven testing, the verification of a software system based on a model of the system. Depending on the respective approach, the tested aspects of a system and its operating environment are represented by models that support the automatic generation of test cases and the allow an automated assessment of the success of the each generated tests. We discuss model-driven testing in Section 2.3 and relate it to MTCC.

2.1 Roles and Properties of Models

The term *model* has a number of different meanings, depending on the context in which its is used. Figure 2.1 gives an overview of 14 different definitions for the term model taken from the online version of the Merriam-Webster Dictionary[1].

For software engineering, the definitions of model as *structural design* and *a system of postulates, data, and inferences presented as a mathematical description of an entity or state of affairs; also : a computer simulation based on such a system* are of particular relevance as they allude to the prescriptive and descriptive aspects of modeling [Béz05].

Main Entry: **¹mod·el**
Pronunciation: \ˈmä-dᵊl\
Function: *noun*
Etymology: Middle French *modelle,* from Old Italian *modello,* from Vulgar Latin **modellus,* from Latin *modulus* small measure, from *modus*
Date: 1575

1 *obsolete* : a set of plans for a building
2 *dialect British* : COPY, IMAGE
3 : structural design <a home on the *model* of an old farmhouse>
4 : a usually miniature representation of something; *also* : a pattern of something to be made
5 : an example for imitation or emulation
6 : a person or thing that serves as a pattern for an artist; *especially* : one who poses for an artist
7 : ARCHETYPE
8 : an organism whose appearance a mimic imitates
9 : one who is employed to display clothes or other merchandise
10 a : a type or design of clothing **b** : a type or design of product (as a car)
11 : a description or analogy used to help visualize something (as an atom) that cannot be directly observed
12 : a system of postulates, data, and inferences presented as a mathematical description of an entity or state of affairs; *also* : a computer simulation based on such a system <climate *models*>
13 : VERSION 3
14 : ANIMAL MODEL

Figure 2.1: Definition of *model* in the Merriam-Webster Dictionary

---

[1] http://www.merriam-webster.com/dictionary/model

2.1.1 Attributes of Models

Greenfield [GS04] defines a model in the context of software development as follows: *"... a model is an abstract description of software that hides information about some aspects of the software to present a simpler description of others..."*. Evans [Eva04] gives the following definition: *"A system of abstractions that describes selected aspects of a domain and can be used to solve problems related to that domain"* where domain is defined as follows: *"A sphere of knowledge, influence, or activity."*. Stahl [SV06] define the meaning of a model such: *"A model is a abstract representation of structure, function or behavior of a system."*

According to Stachowiak [Sta73], a model is characterized by three properties (translation by the author):

- **Representation property** *Models are reproductions or delineation of natural or artificial objects*
- **Abstraction property** *Models do not capture all attributes of represented original but only those that appear relevant to the creators and/or users of the model*
- **Pragmatic property** *Models are not always unambiguously assigned to their original. They fulfill a substitution role a) for certain subjects, b) for certain periods of time and c) under constraints on certain mental or physical operations*

2.1.2 Types of Models

A common aspect of all definitions is the concept of abstraction. For this work we define models as abstractions with a purpose [Lud03]. In the following, we further differentiate models by two characteristics:

- Models may be formal or informal. We consider a model formal when it is represented in a way that allows the automatic processing or transformation of the model for its intended purpose. The structure and semantics of a formal model are defined by its meta model.

  Modeling languages like the UML can be used both formally and informally. When used formally in the context of the MDA [Bro04], the MOF [ALP05] serves as a meta model for the UML. Besides its use for formal modeling, the UML can also be used as a sketch in order to outline the structure of systems.

Whether a modeling language is formal or informal is independent from the question whether it uses a graphical or textual representation [HR04].

- Models can be used either prescriptively or descriptively. Prescriptive models are used for the design of systems that are yet to be constructed, descriptive models represent an system that already exists [Béz05].

From the definition of a model as an abstraction with a purpose follows that the subject represented by the model and the goal of modeling process must be considered in every modeling situation. The subject and the purpose of a model define the domain that is covered by the model. This domain determines the criteria that stipulate which aspects of reality have to be representable by the model at what level of detail.

2.1.2.1 Meta Model

A meta model defines the structure of a model as well as its semantics. The semantics that can be expressed are specific for a domain, therefore, the meta model is a representation of that domain. Figure 2.2 examines the relationship between formal models, their domain and the various concepts that define a meta model. Models are expressed in a domain specific language (DSL). A DSL defines a concrete syntax or formal notation used to express concrete models. The abstract syntax defines the basis concepts available for model construction, it is determined by static semantics of the meta model.

Notwithstanding the existence of a multitude of standards and approaches for meta modeling [KK02, BSM^{+}03, ALP05], meta models and their use are an active field of research. This is also true for the integration of different meta models [ES06, BFFR06] and the representation of constraints on meta models [CT06]. We argue that neither research nor practice have yet succeeded in establishing an universal meta modeling standard.

Meta models are related to ontologies [SK06]. While ontologies aim at the presentation of knowledge, the purpose of meta models is the definition of a formal language.

Figure 2.2 illustrates one important property of models: The separation of the representation of a model, its concrete syntax, from its semantics. This property allows the use of multiple representations for one meta model, for instance the textual representation of UML [Spi03] or — in the case of MTCC — the representation of the feature model through a graphical editor.

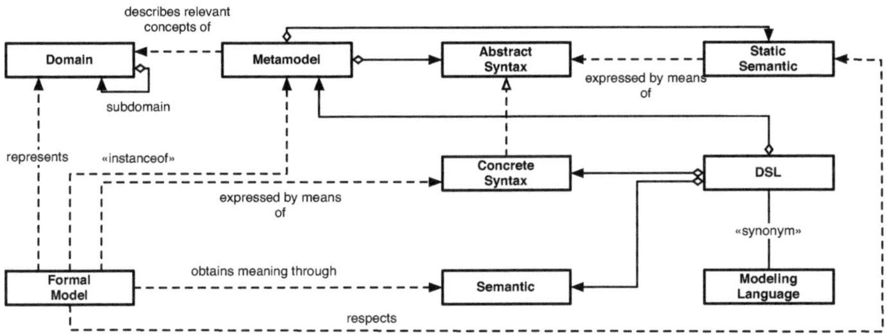

Figure 2.2: Concepts of model-driven software development [SV06]

2.2 Model-Driven Software Development

The purpose of model-driven software development is to close the gap between the problem domain and the solution domain [SV06]. Models abstract from the technical details of an implementation and facilitate the use of concepts from the subject domain not only in requirements engineering and design, but also during the implementation of a system.

France [FR07] express this goal as follows: *"current research in the area of model driven engineering (MDE) is primarily concerned with reducing the gap between problem and software implementation domains through the use of technologies that support systematic transformation of problem-level abstractions to software implementations."*

The stated goals and claims associated with model-driven software development, in particular the MDA approach, have been the subject of critique [McN03, Tho04]. We argue that potential deficits of specific approaches do not challenge the effectiveness of model-driven software development in general but rather that the methods and tools used for any software project must consider the individual requirements of each project.

In the following, we introduce multiple approaches to model-driven software development and relate them to the MTCC. We discuss the transformation of models into implementations. Furthermore we explain the relationship of model-driven development and programming languages with a particular focus on domain specific languages [vDKV00]. We present a summary of relevant approaches to model-driven development, a more detailed comparison can be found in [Völ06a].

2.2.1 Model Driven Architecture

The Model Driven Architecture (MDA) is a standard for model-driven development propagated by the Object Management Group [MSU04, Uhl06].

One defining aspect of the MDA is the successive refinement of models [SV06]. A Platform Independent Model (PIM) represents functional requirements in a formal notation. The PIM is used as the basis for one or more Platform Specific Models (PSM). Each PSM is an abstract representation specific for one aspect of the implementation, for instance, the database layer or the deployment architecture. In order to create an implementation, each PSM undergoes one or more transformation steps, turning the PSM into successively more detailed models and finally implementation artifacts. The MDA emphasizes model-to-model transformations in the software construction process, as the description above illustrates.

We argue that the MTCC approach does not benefit from the advantages offered by the MDA approach because the increase in complexity that the adoption of the MDA would cause is not worthwhile.

2.2.2 Architecture-Centric Model-Driven Software Development

Architecture-Centric model-driven Software Development (AC-MDSD) is an approach introduced by Stahl and Voelter [SV06]. The purpose of AC-MDSD is the *"holistic automation during the development of infrastructure code respectively the minimization of redundant, technical code in software development"*. A concept central to the approach is the platform, the total of all parts of the infrastructure that are used for the implementation. Compared to the MDA, the focus of AC-MDSD is more on the abstraction of technical detail than on the complete representation and implementation of a system with models. We argue that this approach is more pragmatic than the vision offered by the MDA. MTCC shares the goal of AC-MDSD to abstract from infrastructure code; in the case of MTCC, test code and code used to interface with the testee.

2.2.3 Generative Programming

With Generative Programming, Czarnecki [CE00] introduced an approach that aims at creating complete implementations of software systems, optimized for specific requirements, based on a library of implementation artifacts and a formal representation of available configurations. Czarnecki describes the approach as follows: *"Generative software development is a system-family approach, which focuses on automating the creation of system-family members: a given system can be automatically generated from a specification written in one or more textual or graphical domain-specific languages"* [Cza05]. Like Generative Programming, MTCC considers families of systems, more specific families of tests.

2.2.4 Generation of Implementations

In order to facilitate the creation of implementation artifacts from models, a transformation step is necessary [CLW06]. With the execution of one or more transformation steps, implementation details and models are integrated and successively more detailed models are created. A transformation step can either be a Model2Code transformation or a Model2Model transformation [SV06]. While a Model2Model transformation generally starts with an input model and generates another more detailed, model, Model2Code transformations generate an implementation artifact, usually program code. A number of different model transformation approaches exist that address different goals [CH03] and vary in the complexity and suitability for complex transformation scenarios. The test cases generated in the MTCC exhibit only minimal complexity. Accordingly, we limit our consideration of possible transformation strategies to template-based [Her03, Voe03] code generation.

2.2.5 Models and Programming Languages

Model-driven development shares a significant number of concepts with programming languages and compiler technology. Considerable parallels exist between model transformation and compilation — both processes create implementations from abstract representations. The concepts used by programming languages to describe both data and control flow [Sco05] and the rules that govern their use and possible combinations in a program correspond to a meta model.

Domain specific languages [vDKV00, TK04] exist for a wide variety of application domains [MHS05, Hud96]. Such languages have the same purpose as models in MTCC expressing the concepts of a specific application domain — they implement problem-specific abstractions [GKR+06]. Implementation techniques for DSLs are similar to those used in model-driven development [Völ07, Völ06c], especially for approaches that use code generation.

Another approach that serves to illustrate the close relation between model-driven development and programming languages is Languages Oriented Programming. Fowler gives an overview of one such approach [Fow05c, Fow05b, Fow05a], the MPS environment [Dmi04]. MPS aims to facilitate the integrated development of a meta model, an editor for modeling, and transformations for implementing DSLs. The intentional software approach is similar in its goals [Sim95, SCC06].

MTCC uses a graphical editor to present models to domain experts. The XML-based representation of MTCC models can be considered a domain specific language for system and test description.

2.3 Model-Driven Testing

Testing is always Model-Driven in the sense that it is based on a fault model. A fault model considers a system with regards to its likely faults. Binder gives the following definition of a fault model *"(A Fault Model) identifies relationships and components of the system under test that are most likely to have faults. It may be based on common sense, experience, suspicion, analysis, or experiment..."* [Bin99].

Similar to models used in the construction of systems, a fault model does not have to be formal to be useful for testing, it may take the form of a list containing the functionality of a system to be tested. Formal models that are suitable as a basis for automatic testing are called testable [Bin99]. The use of testable models for the automatic verification of software is the subject of model-driven testing.

Figure 2.3 displays an example of a model-driven testing process and the concepts that are relevant during the phases of the process.

- In a first step, a model that can be utilized for the automatic generation of tests is realized based on the requirements of the examined system. The model represents those aspects of the behavior and state of the testee that are relevant for testing. For any possible input to the system, it provides the outputs expected from the system. The models serve as an oracle.

- In the Test Case Specification phase, criteria for the selection of tests are defined and formalized. Criteria include the test coverage by different metrics as well as random selection of tests and the use of previously recorded user interactions with the system. The Test Case Specification is a formalization of this criteria that facilitates the automatic selection of tests cases.

- Tests are generated and executed on the SUT. The adaptor in Figure 2.3 corresponds to the concept of the test runner in MTCC. The results of all executed tests are recorded.

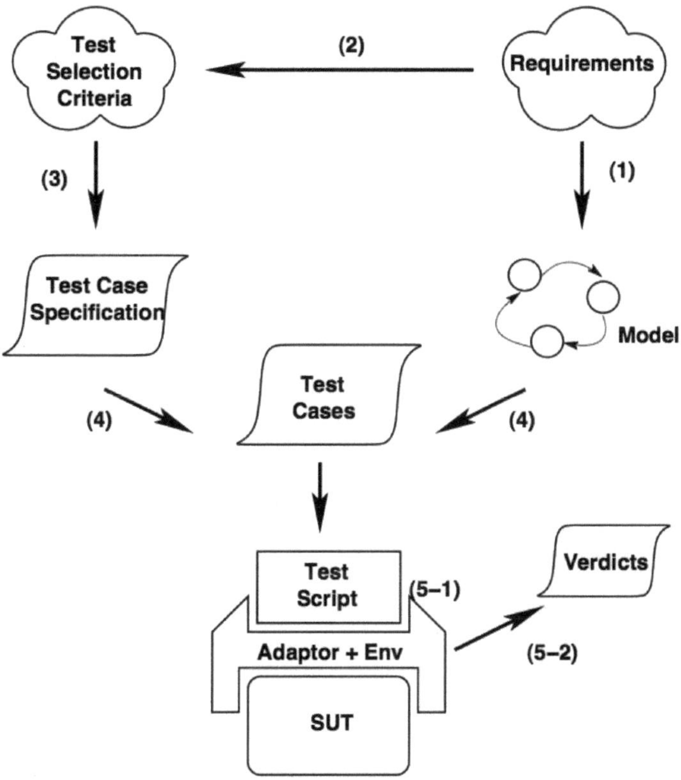

Figure 2.3: A process for model-driven testing [UPL05]

2.3.1 Types of Model-Driven Testing

As in model-driven development, many approaches to model-driven testing exist. A criterion by which different approaches can be differentiated from MTCC is *Redundancy* [UPL05]. Redundancy differentiates between approaches that reuse models created for the construction of a system from those that use specialized models created only for testing. Bertolino uses the term test-based modeling [Ber07] for the latter group.

Approaches that aim at reusing or extending models or modeling tools created for the construction of software are limited by the modeling languages used in this process, often the UML. As a consequence, there is a great number of approaches to model-driven testing that are based on the UML [CCD04, BLC05, SS08, LJX+04, KWD05, HVR04].

One application of the UML for testing is the UML2 Testing profile [SDGR03, BDG+03, Dai04]. The purpose of the U2TP is to provide an UML profile that adds test-relevant concepts to the UML2 core language and thus allows the modeling of tests using UML specific tooling. The U2TP can be regarded as model-driven development applied to the testing domain. By envisioning intellectual test specification, the U2TP bears similarities to MTCC, but differs in the modeling language and the support for modeling by domain experts.

Test-based modeling approaches are not limited to any particular modeling language. Utting [UPL05] present an overview over approaches to model-driven testing and the models used in these approaches. Finite State Machines [EFW01, AOA05], decision trees and decision tables are frequently used as models for testing purposes [Bin99].

2.3.2 Relation to MTCC

Figure 2.4 displays a taxonomy of model-driven testing that classifies approaches to model-driven testing along seven dimensions. Not all dimensions defined by the taxonomy can be applied to MTCC, in the following we only discuss the relevant classification criteria.

Subject MTCC models represent a testee from a system family of testees. The test-relevant Services of a testee as well as the Test Steps that exercise and verify these Services are modeled.

Redundancy The models constructed for the MTCC are only used for testing.

On/Offline MTCC generates and executes test cases offline.

Test Selection Criteria MTCC envision test case construction by domain experts. Domain experts select and implement those test that they consider relevant based on their knowledge of the requirements of the system.

2.4 Chapter Summary

A number of different meanings exist for the term *model*. In this work, a model is a formal abstraction that serves a specific purpose. A model is considered formal if its structure and semantics conform to an implicit or explicit meta-model.

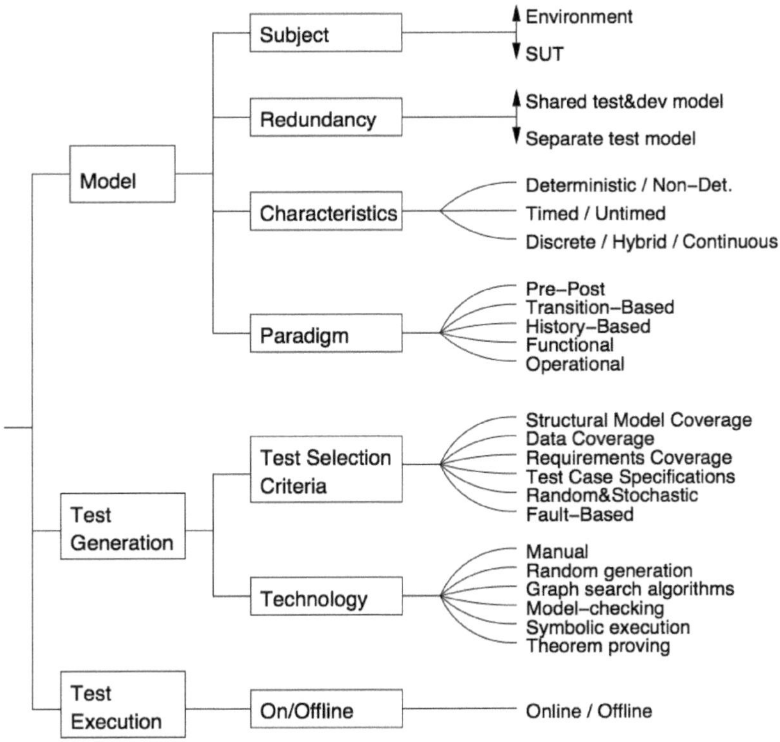

Figure 2.4: A taxonomy of model-driven testing [UPL05]

Model-Driven Software Development attempts to solve the *Problem Implementation Gap* by using formal models that represent concept of the application domain. Applications are modeled and the resulting models are used, optionally after further refinements, to generate implementation artifacts. MTCC is a model-driven approach that uses models to represent SUTs and tests that exercise a SUT.

Model-Driven Testing uses formal models of a SUT and optionally the environment of the SUT to generate tests and to assess the success or failure of these tests. MTCC is closely related to model-driven testing approaches but differs in that tests are not generated but constructed by domain experts.

3 System Families and Variability

In this chapter we discuss system families [Par76] and their relation to software product-lines [WL99]. We introduce feature models [CHE05c] and examine to which degree approaches to Software Product Line Testing [TTK04, PM06] can be applied to MTCC.

Section 3.1 discusses the properties of system families as the object of investigation in MTCC. . A system family is a group of programs who are better discussed in terms of their commonalities than their differences [Par76]. The members of a system family can be programs within the same application domain that fulfill the same tasks but were developed in separation. Alternatively, the members of a system family can be variants or versions of a common product. A Software Product Line [WL99, CN01] is a number of related products which are derived from a common set of core assets. The focus of software product-line engineering is on the organized reuse and adaption of implementation artifacts [Mei06a] and the systematic management of variability [BC05].

One means to formally express the variability in a product-line are feature models [KCH+90] as introduced in Section 3.2. MTCC utilizes feature models both to represent the functionality and properties of a testee and to describe the possible parameterization of Test Steps. Feature models facilitate the representation of the supported features and combinations of features of the members of a system family in an formal, hierarchical way. Feature models define operations that allow derivations of a feature model that represents a more constrained variability. Specialization operations perform local transformations on the feature model that result in a more specific feature model. A fully specialized feature model is a configuration [CHE05b, CHE05c].

From the definition of a system family follows that the requirement exhibited by its members overlap to a certain degree. This leads to the question how tests for the common requirements of systems can be shared among systems. MTCC's primarily means of support for test reuse are abstract test models that can be transfered between systems. Section 3.3 discusses the testing of system families and product-lines and discusses the role of MTCC in this context.

3.1 Variability in Software Engineering

The increasing complexity of software systems and the associated increase in cost necessitates the reuse of implementation artifacts. One context for reuse that is of particular interest to this work are program families or system families[1] [Par76]. Parnas defines program families as follows: *Program families are defined as sets of programs whose common properties are so extensive that it is advantageous to study the common properties of the programs before analyzing individual members*

Different kinds of system families exist, depending on the relationship of the members of the product-line:

- System variants are members of a system family that were adapted for different markets or customers. Variants may differ in their provided features or may be adapted to the individual requirements of customers.
- Product versions are systems where a newer version of a software builds on an older version. Since it cannot be assumed that every customer of a software system can always use the most recent version, different versions have to be maintained in parallel.
- An existing system and its replacement are members of a software family when both systems share no parts of their implementation but are used for the same tasks and are subject to the same requirements.

Despite the existence of various approaches to facilitate reuse in software families, the degree of reuse in software development is still unsatisfactory, reuse in practice is often opportunistic [Mei06a]. In order to support reuse within system families, a number of requirements must be meet:

- The variability [GBS01, BC05] in the system family, meaning the range of common and different features that can exist for the members of a product-line, has to be captured and formalized. Domain analysis [Ara89, Ara94] is employed in order to identify and capture the relevant concepts of the domain that are subsequently formally expressed as feature models [KCH$^+$90, CHE05c]. We call the set of all systems that are covered by the variability of the system family the scope of the system family.

[1] We use the term system family instead of the term program family used by Parnas since we take the position that it better reflects current usage of the terms system and program.

- Processes and methods must be established that, based on requirements for a system to be build, support a statement on whether the scope of the product-line covers these requirements. If the system is within the scope of the product-line, a configuration must be created that describes the particular features of the systems in terms of the variability of the system family.
- The realization of all configurations that can be expressed within the scope of the system family has to be possible. Tooling and core assets must exist that allow the fully automatic implementation of a product from the previously build abstract configuration.

A system family for which all the above requirements are fulfilled, is called a Software Product Line (SPL) [CN01, WL99]. Meister [Mei06a] defines a SPL as follows: *"A product family is called an SPL if it was explicitly designed for the reuse of common elements and the tractable reuse realization of varying elements."*

In Software Product Line Engineering, work on concrete products is done in parallel with work on the artifacts of the product-line infrastructure. Figure 3.1 illustrates this point.

- In Product-engineering, the requirements for a specific system are captured. Based on these requirements a product configuration is prepared and as many features of the product based on the core assets of the Product Line as possible are realized. Features that cannot be automatically realized, are implemented by hand and form the basis for new requirements to the SPL.
- SPL engineering addresses the product-line infrastructure itself. SPL-scoping defines the scope of the SPL, the domain considered by the product-line and the set of products that can be realized by the product-line. Once the scope of a SPL is defined, the set of *Core Assets* [BC05] used to realize members of the product-line are first established [BC05]. core assets for product realization can be implemented based on a number of different approaches. Potential implementations include techniques from model-driven development [VG07, LGR06, CA05], the use of component technology [KMA03], approaches based on programming languages [PM02] and frames [PM03].

3.1.1 Analysis and Scoping of a Product Line

SPL-Scoping, the part of SPL-analysis that addresses the definition of the extent of the domain covered and the concepts within the domain, is a prerequisite for

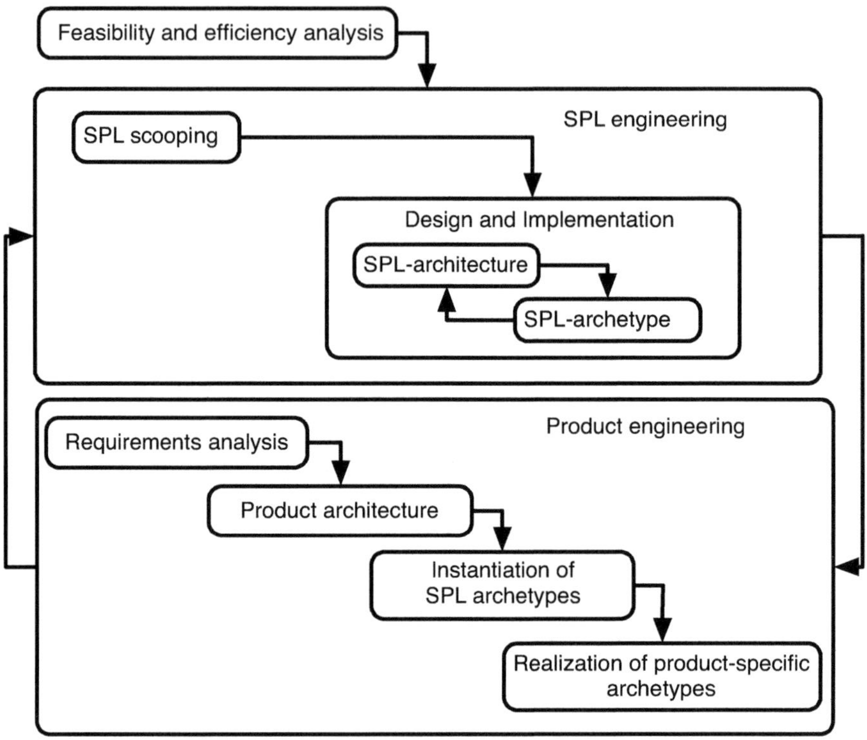

Figure 3.1: SPL development process [Mei06a]

the establishment of a SPL. SPL-scoping is an adaption of the earlier concept of domain analysis that differs in focusing not on the properties of an abstract domain but on the construction of concrete products in the domain [Mei06b, DS99].

An important aspect of product-line engineering is the identification of the relevant features. Czarnecki [CE00] give two conflicting definitions of the term feature:

- *An end-user-visible characteristic of a system.*
- *A distinguishable characteristic of a concept (e.g., system, component, and so on) that is relevant to some stakeholder of the concept.*

For this work we use the definition that originates from the FODA [KCH+90] approach of feature as *'user-visible aspects or characteristics of the domain'* We argue that this view on features better represents the external view of a system taken in acceptance testing.

Multiple approaches for domain analysis and SPL-analysis are defined software engineering [BFK+99, BC05, CE00, Nei89, CN01, WL99, Mat04]. Each approach defines information sources used in the analysis process and defines the methods and notations used to formalize the knowledge gained in the analysis. Figure 3.2 gives an overview of possible sources of information for domain analysis. In order to identify test-relevant functionality, MTCC considers the systems to be tested in the system family as well as the test regime in use and the requirements formulated by the domain experts. The latter is of particular importance in situations where no test for some functionality of the testees exists.

With the exception of feature models, the details of domain analysis, such as the process employed and the artifacts used, are not discussed in detail. MTCC does not assume the utilization of any particular approach for the capture and formalization of the concepts and features of a domain. The exclusion of these subjects from MTCC is due to two reasons: First, the contribution does not lie in the field of analyzing but in the application of domain knowledge, and second, the analysis necessary for MTCC as it is implemented for this work is simplified by the fact that the system family and all systems to be tested already exists. The domain analysis process is therefore not subject of this work.

Feature-Products Maps [DS99] provide an overview, which products of a system family support which features. Every feature is correlated with every product in the system family. The features themselves and thus the variability of the system family are represented by feature models.

3.1.2 Variability and System Family Testing

We consider two aspects of the variability of a system family of particular importance for MTCC:

- MTCC does not assume that the testees considered are members of a software product line. Instead, MTCC is based on the premise that the testees form a system family defined by a common set of test-relevant features. MTCC has the characteristics of a product-line approach for the construction of tests. MTCC Core Assets are the models used to represent systems and tests as well as the editor and the subsystems responsible for model transformation and test code generation.
- In this work, we only apply MTCC to an already existing system family of Digital Libraries. This constraint allows for simplifications in the domain

Source	Advantages	Disadvantages
Textbooks	• Good source of domain knowledge, theories, methods, techniques, models	• Reflects only specific author's views • May use idealized or biased models
Standards	• Represents standard reference model for domain	• Model may not be current with new technology
Existing Applications	• Most important source of domain knowledge • Can be directly used to determine user-visible features • Requirements documents available for domain model • Detailed design & source code show architectures	• Cost of analyzing many systems is high
Domain Experts	• Can provide contextual/rationale information unavailable elsewhere • Can be consultant during DA, validator of products afterwards	• Experts have different areas of expertise; several experts may be needed

Figure 3.2: Information sources for domain analysis [KCH$^+$90]

analysis phase that cannot be assumed if the systems in the system family are still in the development process.

3.2 Feature Modeling

The purpose of feature models is the representation of the variability in a system family. A feature model organizes the features of a system and the dependencies of the features in a tree structure. Specialization steps are operations that support the refinement of the variability and thus the specification of concrete systems in terms of their features. MTCC uses feature models for the representation of the Services provided by testees as well as for the representation of the parameterization of tests.

3.2.1 Structure of Feature Models

A feature model is a tree structure that facilitates the hierarchical description of the features of a system. Features within a feature model can represent system-subsystem relationships, choices among different functions that a system can provide or abstract, non-functional requirements. Figure 3.3 displays a simple feature model in the notation introduced by the FODA [KCH+90] approach. The feature model in the example represents the features of a car as follows:

- Every feature is the composition of its children. The feature Car is defined by the lower-level features Transmission, Horsepower and Air conditioning.
- Features can be mandatory or optional. While every car must have a transmission, Air Conditioning is optional and can be de-selected.
- Features can be alternatives. The Car represented in the feature model either has an automatic or a manual transmission, but never both.
- Features can hold numerical values, for example, the feature Horsepower.

Alternatives like the one between automatic and manual transmission in Figure 3.3 are a special case of a feature group. Besides alternatives that allow only one of the features from the feature group to be selected, FODA also defines feature groups that allow the selection of an arbitrary number of features.

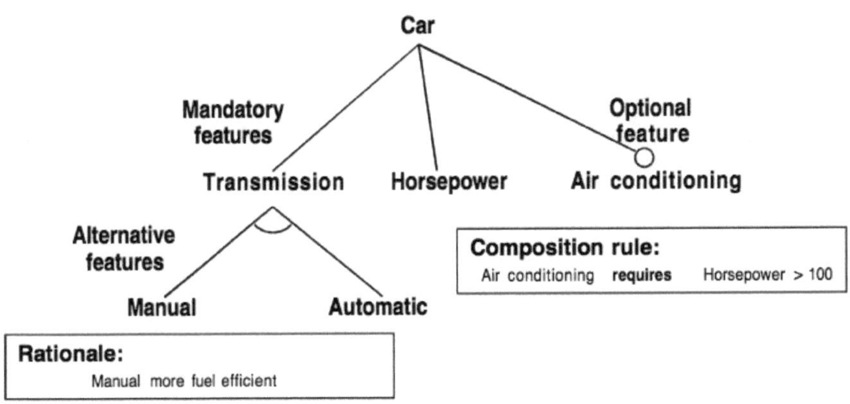

Figure 3.3: Example for a FODA Feature Diagram [KCH+90]

Not all relevant aspects for the configuration can be represented by the hierarchical structure of the feature model alone, rationales and composition rules are employed to add extra information to a feature model. A rational gives a human-readable reason that can serve as a basis for the selection or de-selection of a feature. Composition rules describe under which conditions certain features can be selected or combined. A feature diagram is a set of feature models together with their rationales and composition rules. MTCC does not use rationales or composition rules. We argue that the complexity that these rules add to feature models is not justified in the context of MTCC, primarily because of the descriptive, fine grained nature of the feature models used in MTCC.

An important property of feature models is the abstraction of implementation details. While automatic approaches exist that map features to implementation artifacts [CA05], the overall level of abstraction provided by feature diagrams is higher than is the case for modeling approaches that represent high-level implementation artifacts such as UML class diagrams.

Based on the notation and the semantics for feature models defined for FODA, a number of different extensions were developed [HST07]. The feature models used in MTCC are based on feature models with cardinalities [CHE05b, CHE05c]. Compared to the feature models defined by FODA, feature models with cardinalities exhibit the following extensions.

- **Cardinalities for Features** The cardinality of a feature node describes how often the feature and its children can be included in the feature model. Cardinalities are expressed as an interval [n,m] where n describes the minimal occurrence of a feature and m its maximal occurrence. Mandatory features have a cardinality of [1,1], the cardinality of an optional feature is [0,1].

- **Attributes** Attributes are nodes in a feature model that can hold numerical or textual information.

- **References** Reference nodes support the inclusion of a referenced feature model or a subtree of a feature model in place of the reference node.

3.2.2 Configuration and Specialization of Feature Models

A feature model describes the variability within a system family in terms of all its potential configurations. A configuration is a feature model without any variability, the feature model is completely specialized.

Czarnecki [CHE05c] describe the relation of a feature model and its configuration as follows: *'The relationship between a feature diagram and a configuration is comparable to the one between a class and its instance in object-oriented programming'*. The transformation of a feature model in a completely specialized form is called the configuration process [CHE05c]. The configuration process is defined by the application of a series of specialization steps. Each specialization step creates a new feature model with less variability than the initial feature model.

Seven specialization steps are defined for feature models with cardinalities. Figure 3.4 illustrates their application on feature models:

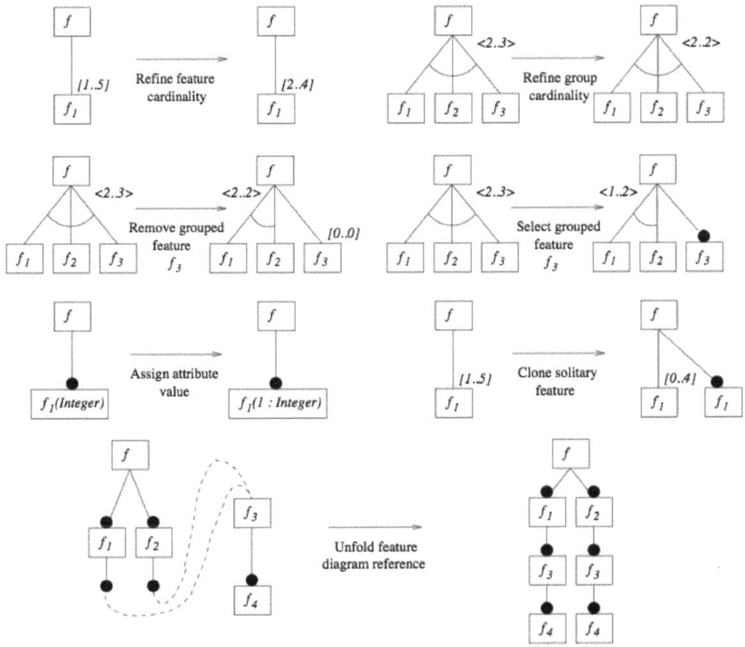

Figure 3.4: Specialization steps [CHE05b]

Refinement of feature cardinalities The refinement of feature cardinalities adjusts either the minimal or the maximal occurrence of a feature in such a way that the range between the minimum and and maximum is smaller than in the initial feature model.

Refinement of group cardinalities This specialization step narrows the interval that describes the minimal and the maximal number of selectable features in a feature group.

De-selection of features from a feature group This specialization step sets the cardinality of one child of a feature group to [0,0], the node can no longer be included in a configuration. By de-selecting the node, the number of maximal children of the feature group is lowered by one.

Selection of one feature from a feature group This specialization step sets the cardinality of one child node of a feature group to the interval [1,1]. By applying the specialization step, the minimal as well as the maximal number of children for the feature group are lowered by one.

Assignment of attribute values This specialization step sets the value of an attribute. If a type system for attributes is used, as is the case in the situation illustrated in Figure 3.4, the legal values are constrained by the type system.

Cloning of solitary feature nodes As the name suggests, the cloning specialization step duplicates a feature node which have maximum cardinality greater than 2. The cardinality of the cloned feature is [1,1], the minimum and maximum occurrence of the cloned feature node are lowered by one.

Unfolding of feature reference nodes When a feature reference is unfolded, the referenced features are included in place of the feature reference.

As described by Czarnecki et al [CHE05b] for the Staged Configuration Process, the configuration of feature models in MTCC is done in a series of incremental phases. First, a representation of a single system is derived from a description at the system family level, then references from Test Step models to the system model are unfolded, and finally the feature models of Test Steps are configured.

Some restrictions apply to the implementation used in MTCC for feature models with cardinalities. Solitary features are limited to the cardinalities [0,1], [0,0] and [1,1], refinements of cardinality and cloning of features are not supported.

3.3 Testing of System Families

A number of approaches address testing in the context of system families or software product-lines. Compared to the testing of solitary systems, such methods face a number of specific challenges:

- Ideally, the development of tests and the infrastructure needed for test execution should be based on a set of core assets. The testing of single systems corresponds to the product engineering activity [PM06] in the context of a product-line while work on common test assets is related to Software Product Line Engineering.

- The features that are available for different systems under test within a product-line or system family have to be considered in test selection. This requirement has two aspects: Each system can only be exercised when it exhibits all features necessary for the execution of a test and each feature of a system should be covered by a test [GLRW04, MSM04]. We call this requirement the test selection challenge.

- Test runners and interfaces must exist that support the execution of tests for all systems within the system family. In a SPL approach, the implementation artifact must support testing [KM06].

Software Product Line Testing (SPLT) is an area of intensive research [TTK04]. A significant difference between SPLT approaches and MTCC is that MTCC does not assume that the testees are members of a SPL and are built from a common set of core assets.

A significant difference between testing in the context of software product lines and software product families that are not based on a common set of core assets is that no commonalities in the implementation or the interfaces of the member of the system family can be assumed. Furthermore, no formalization of the variability of the different members of the system family will be available for the members of the system family — the members of the system family were not configured.

As we already pointed out, MTCC does not assume that testees are members of a software product-line. As a consequence, neither the existence of core assets nor a formal representation of the variability of the product-line can be assumed. Since reuse of existing assets is not possible, the identification and capture of variability in the context of MTCC is done for the first time and only for the purpose of testing. We argue that while the reuse of product line assets for testing would be desirable, the use of models dedicated to testing simplifies the modeling process, thus the construction of the MTCC test models is far less costly than the creation of models for software construction.

MTCC does not aim at the fully automatic testing of product-lines. Since domain experts select the systems of the system family to be tested as well as the features of a system to be tested, one aspect of the test selection challenge introduced above does not apply. The other aspect of the test selection challenge, to ensure that only tests supported by the features of system are specified for that system, is addressed by the MTCC editor and its approach to test reuse.

MTCC addresses the remaining requirements and particularities of testing in a system family context as follows:

- The core assets of the MTCC approach, the infrastructure necessary for testing all members of a system family, are strictly separated from artifacts that are specific to the testing of a single system. Specifically, the editor, the Test

Generator and the models used to represent the system family at the domain level are used for all systems tested with MTCC.

- The MTCC testing process is organized in two parallel activities: The derivation of system-specific tests and the work on domain models and testing core assets.
- Tests constructed in MTCC, in the form of Test Configurations for one specific member of the system family, are potentially reusable for multiple members of the system family. A test is reusable for all systems which features support the execution of the test.

While we do not address this aspect in this work, the formalization of the variability within a system family and the representation of system family members as configurations facilitates other means of quality assurance than automated acceptance testing. For instance, with a suitable detailed description of the features of individual systems and an infrastructure that adapts the different interface of testees, Differential Testing [McK98, ES07] can use the behavior of one system as an oracle for other systems under test.

3.4 Chapter Summary

MTCC is a testing approach for system families. A system family is a set of systems that are defined by their shared properties, not their differences. While Software Product Lines are based on the systematic reuse and adaption of a common set of Core Assets, system families are generally not the product of systematic reuse. MTCC addresses the testing of system families and therefore does not assume the existence of common infrastructure or artifacts.

A central aspect in working with system families and product lines is the systematic treatment and representation of the variability. Feature models are a means to represent the features that exist in a system family and their relationships. MTCC uses feature models with cardinalities to represent the test-relevant features in a system family and the tests that exercise these features.

4 Information Retrieval and Digital Libraries

In this work, we apply MTCC to a system family of Digital Libraries, specifically to web portals for scientific information. Digital Libraries are a class of Information Retrieval systems; the functional quality of such systems is defined by their ability to fulfill the information need of a user and present him or her with relevant documents for a search.

The vagueness inherent to the Information Retrieval task makes the participation of domain experts in the testing process particularly important. Section 4.1 introduces the concepts and models relevant to Information Retrieval, special focus is given to evaluation measures for Information Retrieval and the notions of quality in IR.

Section 4.2 gives an overview of the application domain of Digital Libraries. We introduce basic usage scenarios for this class of systems and discuss the meaning of quality for a Digital Library. We also outline the treatment of structural and semantic heterogeneity as a Service that, while it supports users in searching, also raises the complexity of Digital Libraries and thus necessitates testing.

4.1 Information Retrieval

The field of Information Retrieval (IR) addresses the question how the information need of a user concerning a collection of documents can be captured, operationalized, and satisfied. To this purpose, a number of different retrieval models were designed and implemented, for instance the Boolean, the Vector or the Probabilistic Model. These models define representations for information needs as queries and documents as document surrogates. They further provide operations that allow the calculation of the similarity of queries and document surrogates and thus support a statement about the relevance of a document.

Information Retrieval is an empirical field, a number of measures and evaluation methods exists to support the validation of retrieval models as well as retrieval systems.

4.1.1 Purpose of Information Retrieval

The purpose of Information Retrieval can be well illustrated by contrasting IR with data retrieval [BYRN99]. Data retrieval deals with formal queries that support for each document a definite, automatic, unchanging decision whether the document satisfies the criteria defined by the query. The subject of Information Retrieval in contrast is not a formal query but a vague information need. In addition, if documents in data retrieval are highly structured records, for example, taken from a relational data management system, documents in Information Retrieval are either unstructured texts, semi-structured [ABS00] or sparse.

Manning [MRS08] define IR as follow: *"Information Retrieval is finding material (usually documents) of an unstructured nature (usually text) that satisfies an information need from within large collections (usually stored on computers)"*.

Depending on the information need of a user, the document collection and the nature of the retrieval task, and in absence of a decidable model of relevance, IR bears resemblance to the proverbial search for a needle in a haystack. In reality, this comparison does not illustrate the full breadth of different scenarios covered by IR. Beyond the search for a specific needle in a well-defined haystack, [Kol00] lists the following scenarios for IR:

- *a known needle in a known haystack*
- *a known needle in an unknown haystack*
- *a unknown needle in an unknown haystack*
- *any needle in a haystack*
- *the sharpest needle in a haystack*
- *most of the sharpest needles in a haystack*
- *all the needles in a haystack*
- *affirmation of no needles in the haystack*
- *things like needles in any haystack*
- *let me know whenever a new needle shows up*
- *where are the haystacks; and*
- *needles, haystacks — whatever.*

An information need can address specific documents or all documents that have specific attributes. The needed information can be clearly defined — at least in the mind of searcher — or it can be only vaguely outlined. The information can be needed in the present or sometimes in the future. From the user perspective, a document is considered relevant if it satisfies the information need of a user. In order for an IR system to estimate the relevance of a document, a model of the information need must be compared with models of the documents [Coo73]. Figure 4.1 illustrates the relation of an information need and its representation as a query and documents, represented by document surrogates.

From the technical perspective of an IR system, a document is relevant when the document surrogate and the query are similar.

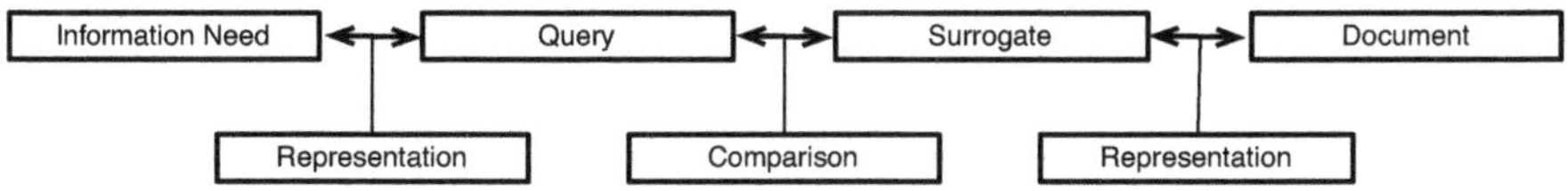

Figure 4.1: Concepts of Information Retrieval [BC87]

The design, implementation and evaluation of retrieval models for the representation and comparison of information needs and documents is a focus of research for Information Retrieval. One aspect of Information Retrieval that is of significance, but can rarely be represented in IR models, is the dynamic nature of an information need. Rather than answering isolated queries, the IR process is an interaction between a user and the IR system in which an information need is refined or changes.

When IR is regarded as a dialogue between a user and a IR system, the support that the IR system can provide for different users, their skills, preferences and backgrounds becomes more important. To meet the requirements that arise from a dialog-centric or user-centric view on IR, the field has developed a number of different interaction models, user interfaces and methods to tackle the use and enhancement of queries and results:

- In ad-hoc retrieval, users formulate queries to a collection of documents. In filtering, documents are added to a collection and are tested for their relevance on a set of predefined queries.

- In search, a user expresses a specific information need in a query and sends this query to the IR system, the systems returns the subset of all documents that are likely relevant to the query. In browsing, the user does not necessarily have an explicit information need, rather, a collection of documents is explored.

Information Retrieval as a field has produced various approaches to support the formulation of information needs and the visualization of results. An overview is presented by Baeza-Yates and Ribeiro-Neto [BYRN99]. Graphical user interfaces exist for the treatment of structural and semantic heterogeneity [Ste03] or for the combined treatment of the usability and design [Eib00].

Methods that facilitate the transformation or enhancement of initial queries have a great impact on the ability of an IR system to return documents that are relevant to the information need of a user. Such methods facilitate the integrated retrieval over collections that use different indexing vocabularies [MP08] or support the consideration of specialized discourse domains in search [Pet06].

Approaches to clustering and automatic classification [Cha03, Wei05] support the enrichment of a result set of documents by further information that support the user in refining his or her information need.

4.1.2 Concepts and Methods of Information Retrieval

The relevance of a document regarding an information need is represented by the similarity of a document surrogate and a query in the context of a retrieval model.

Figure 4.2 gives an overview of retrieval models in IR. Since MTCC takes an external, responsibility-based view of testing, details regarding the theory and implementation of the retrieval model used by the testee are neither available to MTCC nor of interest to acceptance tests. We therefore limit our discussion to the differences between the Boolean retrieval model and retrieval models that support the ranking of results by relevance.

The subjects of retrieval models are not an information need and documents but rather queries and document surrogates.

A common format for document surrogates that is used both by the Boolean and the vector model is the representation as a multi set or bag of words that are included in or relevant to the document.

Depending on the specific transformations that a document undergoes during indexing, the terms in the document surrogate can be identical to those in the document. The terms in the surrogate may also form a subset of the terms in the document, for instance, when stop words are removed from the document; or the surrogate may contain a superset of the original terms, for instance, when synonyms are added to the document. In addition to the removal or addition of terms, transformation processes at the term level such as stemming and lemmatization replace individual terms in documents.

If a document is structured into different fields, for instance the title, author and abstract of a bibliographic record, a document can be represented by a surrogate that contains a multi-set for each field of the document.

Two types of retrieval models can be distinguished: 1) Models that are based on a purely binary model of relevance and 2) vague or graded retrieval models that

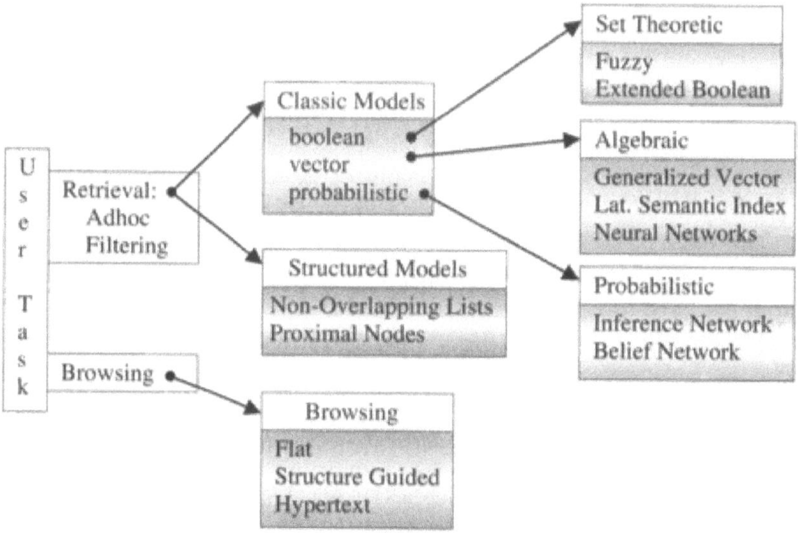

Figure 4.2: Overview of retrieval models [BYRN99]

support the ranking of the documents according to their likely relevance. When a retrieval model with a binary model of relevance is used for search, each document of an IR system is assigned to one of two sets, one containing documents that are relevant to the query and the other containing documents that are not relevant. The result that a system with a binary IR model returns for a query contains all documents from the set of relevant documents without further distinction. Systems based on a graded retrieval model return results sorted by their decreasing probability of relevance, they assign a relevance value to each document.

The Boolean model is based on Boolean algebra: Queries are represented as Boolean expressions. Queries are answered as follows: For each term in a query a set of all documents is built that contains this term, then, unions and intersections of the individual sets are built based on the Boolean operations used in the query. The Boolean retrieval model supports the formulation of complex and precise queries by experts for a retrieval system but is problematic for less experienced users. Not optimally formulated queries can either lead to result sets that are empty or contain so many elements that the user is overwhelmed. A problem inherent to the Boolean retrieval model [Eib00] is that it cannot differentiate between documents that nearly match a query: If a document matches all but one of the terms specified in a conjunctive query, it has the same negative relevance as a document that contains no term at all.

The limitations of the Boolean model were motivating factors in the research of graded retrieval models like the vector model or probabilistic model. The vector model represents both queries and documents as vectors in the vector space whose dimensions are defined by the set of all terms in the information retrieval system. The value of each component of the vector represents the weight of the corresponding term to the document or query.

The approach used to determine the weight of a term has significant impact on the calculation of relevance values by the IR system [SB88]. Basic concepts to calculate term weights is the use of term frequency (tf) and inverse document frequency (idf).

Term frequency is the number of occurrences of a term in a document. The inverse document frequency is defined as the logarithm of the quotient of the number of documents in the IR system and the number of documents in which the term is included.

$$idf_t = \log \frac{N}{df_t} \tag{4.1}$$

Based on the tf, idf and a normalization factor, the weights of all terms for all documents are calculated.

The similarity between a query and a document surrogate is done on the basis of a document vector and a query vector, a frequently used measure is the cosine of the angle between the two vectors:

$$sim(d_1, d_2) = \frac{\vec{V}(d_1)\vec{V}(d_2)}{|\vec{V}(d_1)||\vec{V}(d_2)|} \tag{4.2}$$

The vector model assigns a graded relevance value to each document held by the retrieval system. It is a vague retrieval model in the respect that for queries containing multiple terms, it can also return documents that do not contain all terms. The vector model is better suited for inexperienced users who lack the skill or inclination to formulate complex queries based on the Boolean model. At the same time these properties of the vector model limit its applicability for the formulation of complex and precise queries.

In addition to the Boolean and the vector models, another relevant retrieval model is the probabilistic model [RvRP80] that applies probabilistic theory to the Information Retrieval task and provides graded relevance values for queries. The extended Boolean model [SFW82] combines the support of the Boolean model for formulation of complex queries with the ability to return documents graded by relevance. The combined support for complex queries and graded relevance ranking is a feature of most current IR systems, independent from the internally used retrieval model.

4.1.3 Evaluation for Information Retrieval

Since the effectiveness of an Information Retrieval system is defined as its ability to identify documents relevant to a user's information need and as the relevance of a document with regards to an information need cannot be formally proven, the effectiveness of an IR system must be assessed by empirical means. In order to evaluate an IR system and to compare the effectiveness of different systems, a standardized collection of information needs, documents and in some cases relevance assessments for the documents regarding the information needs are used. We call such a collection an evaluation collection. Using a collection, the evaluation of an IR system is done as follows:

- Every information need included in the corpus is formulated as a query that is then executed on the IR system.
- The documents returned for the query as well as their ranking are recorded.
- The (ranked) list of documents for each query is compared with the relevance assessments included in the corpus. If the corpus does not include assessments for all documents, result documents are intellectually assessed.
- Metrics for the effectiveness of the IR system are calculated based on the relevance assessments. The MTCC approach supports the calculation of Relevance and Precision as explained below

A number of metrics exist for the evaluation of IR systems. The two most basic measures in use are Recall and Precision.

$$Recall = \frac{Number\,of\,relevant\,documents\,in\,the\,result\,set}{Number\,of\,relevant\,documents\,in\,the\,IR\,system} \tag{4.3}$$

$$Precision = \frac{Number\,of\,relevant\,documents\,in\,the\,result\,set}{Number\,of\,documents\,in\,the\,result\,set} \tag{4.4}$$

Recall describes the completeness of a result set in relation to all relevant documents in the IR-system, precision the ratio of relevant documents and documents that were falsely returned as relevant. Since optimizations of an IR system for either recall or precision will generally have a negative impact on the other measure, recall and precision values for a system must always be considered in combination [MRS08].

4.1.3.1 Evaluation of Graded Retrieval Models

As described above, recall and precision are best suited for the evaluation of systems with a binary concept of relevance. Graded retrieval models return a ranked list of documents, for such systems and for large result sets it is not only important that relevant documents are part of the result set but also that they are ranked correctly, since most users will only inspect the first, most highly ranked documents [Bla80, Coo73]. Since recall and precision are based on sets and thus ignore the order of result documents, other metrics are needed for the evaluation of graded IR systems.

One means for the evaluation of ranked result is a Precision-Recall chart. Such a chart displays the precision of a result set at different levels of Recall. Figure 4.3 displays a variant of such a chart, an averaged 11-point precision / recall graph. An advantage of such a representation is that it eases the comparison of results for different evaluations. As illustrated by Figure 4.3, precision falls with increasing recall, the first results returned by the system are therefore the most likely to be relevant by the judgement of the system.

For a number of situations, for instance, when it is likely that a user will only inspect a limited number of documents, regardless of their relevance, the precision value after the first k documents is better suited to evaluate an IR system than precision at a specific recall level. Precision at k and R-Precision are measures applicable for such situations.

As the name implies, precision at k measures the precision of the first k documents of a result list. R-Precision considers the precision after the first k documents known to be relevant of been returned, compared to precision at k, this allows for better comparability between queries with large difference in the number of returned results.

Beyond the measures presented here, a number of specialized measures exist, for example for web-retrieval [Lew07]. Mandl [Man08] discusses the quality of websites for Information Retrieval with respect to automatic assessment of quality.

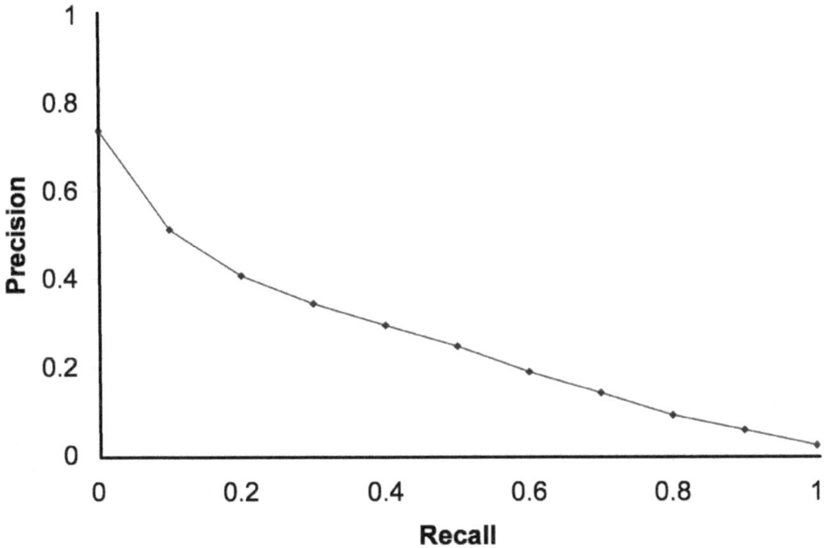

Figure 4.3: Averaged 11-point precision / recall graph [MRS08]

4.1.3.2 Limitations of the Notion of Relevance

The evaluation process used in Information Retrieval and the measures used in that process are based on a number of simplifications that have to be considered in the interpretation of results. One limitation of the evaluation processes outlined above is the use of binary relevance assessments for the documents in the corpus. While this is a simplification of reality, it does not put evaluations done with this measure under question [MRS08].

A more fundamental simplification is the assumption that the concept of relevance is static for every user at each moment of time. This ignores the fact that, for instance, the relevance of a document changes once this document is known to the user.

A final weakness of the evaluation process is that it is assumed that the information need of a user is fully formed and can be expressed in a single query. This ignores the fact that the Information Retrieval process is a dialog and that the information need of a user is evolving [SEN90].

The evaluation of Information Retrieval as a dialog is not possible with the system-centric evaluations discussed above. A user-centric [Voo02] approach that includes aspects like the usability of a system is needed instead. Since MTCC abstracts from the interface that a user employs to interact with the SUT, is not intended to serve as a tool for a user-centric evaluation.

4.2 Digital Libraries

Digital Libraries are specialized Information Retrieval systems. The digital libraries discussed in this work are topic-specific portals that provide scientific information [MSW05] for specific fields of science. This work investigates a family of on metadata-driven Digital Libraries, primarily for the social sciences. We argue that Digital Libraries outside this system family, for example the ACM[1] or the IEEE Digital Library[2], can also be represented and tested using the MTCC approach.

We use the term portal to indicate that these particular Digital Libraries integrate various different databases but use the term Digital Library for the remainder of this text. Figure 4.4 illustrates the relations of portals, topic-specific portals and databases.

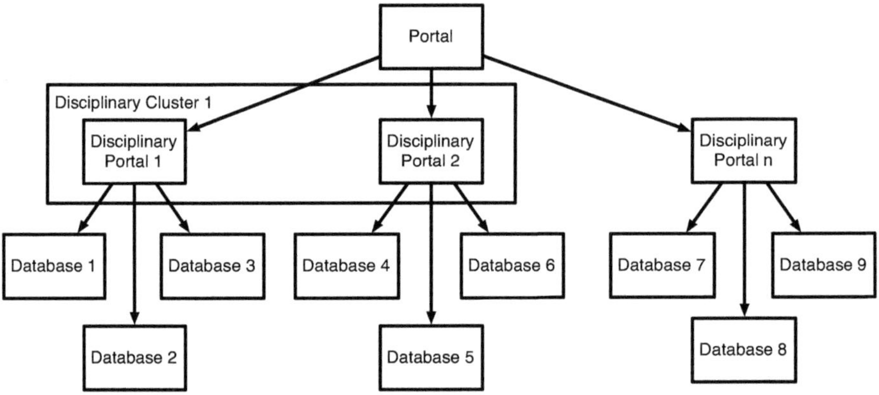

Figure 4.4: Portals, disciplinary portals and databases [MSW05]

The fact that the Digital Libraries discussed here integrate several databases raises the complexity of each system considerably. One reason for this increase in

[1] http://portal.acm.org/
[2] http://http://www2.computer.org/portal/

complexity is the presence of data from a number of different sources. The Digital Libraries discussed here can include the following types of information [May06]:

OPACs Online Public Access Catalogs contain information about the books, periodicals, and other media available in a library. The records of an OPAC usually contain bibliographic data and a subject description in an indexing vocabulary, but do not have an abstract.

Full-text Electronic full-text may be either available at costs or, for instance in the case of Open Access Repositories, for free. While the cost of full-texts is not directly relevant to IR, the treatment of free and at cost resources raises the complexity of a Digital Library.

Topic Guides Topic guides (Fachinformationsführer) describe Internet resources of relevance for a particular field of science, for instance, the homepages of institutions.

Topic Database Topic databases (Fachdatenbanken) are collections of information that supply scientists with research-relevant information. A important aspect of topic databases are bibliographic records, often containing an abstract as well as subject descriptions from a indexing vocabulary.

In addition to these information types, Heinz and Stempfhuber [HS07] list 14 different types of information, called modules, among them expert data bases and event calendars.

The integration of potentially structurally as well as semantically heterogeneous databases in a Digital Library gives rise to a number of challenges concerning the implementation technology, information science, and organizational / legal issues.

One approach to the management of different information types is the Shell Model defined by Krause [Kra96, Kra06a]. As illustrated in Figure 4.5, the Shell Model places different kinds of information in concentric layers around a core of high-quality information. The shell model thus supports the organization of different types and qualities of information by their likely relevance to the user of a digital library.

4.2.1 Structural and Semantic Heterogeneity

Structural and semantic heterogeneity [Ste03] are a consequence of the integration of multiple databases within one Digital Library. Databases are structurally heterogeneous when the schema used to describe information differs. Differences can include different field names, incompatible data types and ranges of legal values

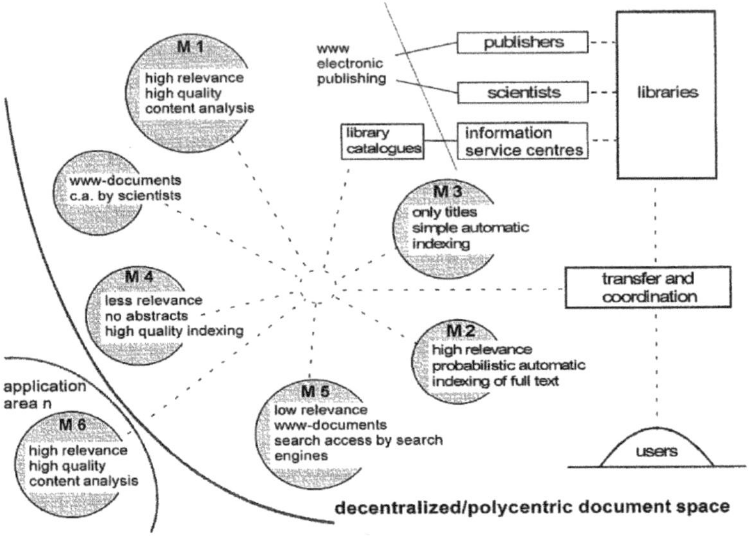

Figure 4.5: Shell Model[Kra06a]

and different granularity used to describe concepts. Structural heterogeneity is a relevant problem outside the domain of Digital Libraries [Has97, Has02, PH04]. Rahm and Bernstein [RB01] give an overview of the challenges imposed by structural heterogeneity and outline some approaches to their solution.

Semantic heterogeneity is caused by the use of different Knowledge Organization Systems (KOS) like thesauri, classifications and subject lists for the different databases. Since a user of a Digital Library cannot be expected to be familiar with multiple KOSs, it is important to make the different KOSs used in a Digital Library interoperable [ZC04]. One means to provide such an operation are cross-concordances [MP08].

Structural and semantic heterogeneity as well as use of methods for treatment of such heterogeneity like cross-concordances increases the complexity of a Digital Library as a technical system. This rise in complexity increases the probability of defects and thus necessitates more testing.

4.2.2 User-Oriented Quality of Digital Libraries

If we define the quality of a Digital Library as the degree to which it satisfies the needs of a user, it becomes obvious that the quality of a Digital Library may be different to each user. Surveys concerning the factors that define the quality of a Digital Library to a user [Thu07] place emphasis both on the importance for a Digital Library to support complex queries and on the usability of the system. In more detail, users have the following requirements for a system [Hei07] (translation by the author):

- *High coverage of potential information sources*
- *Ease of the search process:*
  - *quickly accessible search forms for a simple search*
  - *an uncomplicated way to modify the results of a search that allows the reuse of previous searches.*
- *Simple assessment of the relevance of results:*
  - *by the preview of a document in the list of results*
  - *by sorting the result by different criteria*
- *Simple access to the needed information*
  - *links to document sources*
  - *direct and uncomplicated access to electronic texts*

The significant role of usability is confirmed by Lewandowski [LH07] in the context of web-retrieval. It was also found that advanced functionality for such systems, for example, Boolean operators, were rarely used. While it is unclear, how far these findings apply to the users of Digital Libraries, we argue that differences between available functionality and used functionality are likely to exist for digital libraries as well. In combination with the survey results outlined above we argue that the assessment of the quality of a Digital Library must involve users.

4.2.3 Testing of Digital Libraries with MTCC

In order to test a Digital Library in practice, some simplifications have to be accepted. The following are relevant to testing with MTCC:

- The ability of a Digital Library to provide information relevant to an information need can only be tested using the system-centric evaluation approaches outlined in the previous section. Since MTCC abstracts from the specifics of the graphical user interface, testing of IR as an interactive process involving the information need of a user at a specific point in time is limited.
- Tests can only represent a static information need, the expected behavior of a test executed multiple times cannot change.

Similar to the evaluations of IR effectiveness, testing IR systems is based on simplifications. Figure 4.6 puts the previously discussed concepts of Information Retrieval in relation to the abstractions used in MTCC. A query is a representation of an information need, tests are representations of the requirements for a system. As an IR system can only judge the relevance of a document, tests in MTCC can only verify the testable aspects of a system that are represented by the system model.

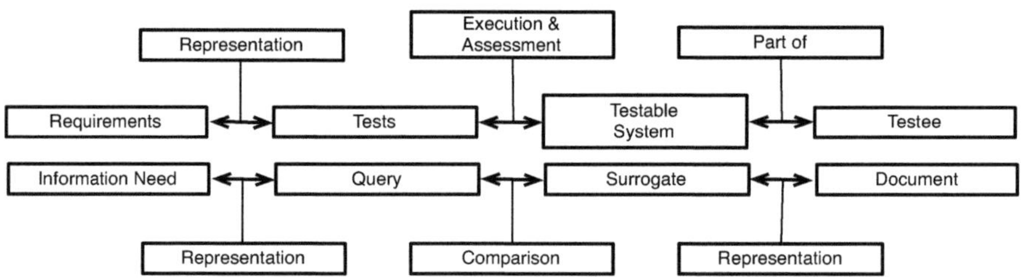

Figure 4.6: Abstractions in Information Retrieval and abstractions in MTCC [BC87]

As discussed in Chapter 1, testing can only improve the quality of a system by finding faults. Applied to the testing of Digital Libraries with MTCC, this means that MTCC must be able to model the Test Steps and Services needed to fulfill Information Retrieval tasks with MTCC. In particular, MTCC must be able to compare the expected and the actual results for a query.

The fact that MTCC is a model-driven approach and thus abstracts from some aspects of tested systems has consequences for the aspects of digital libraries that can be tested with MTCC. The graphical user interface of a system determines what interactions between a user and a IR system are possible and is an important aspect of the user-centric quality of a system. Since MTCC does abstract from all aspects of the user interface, the user interface aspects of a IR system cannot be tested in MTCC. More generally, MTCC does not support testing of any usability-related aspects of a system. MTCC does support the testing of the functional aspects of a IR-system within the constraints of the Cranfield paradigm [Voo02] as well as the testing of the technical functionality and correctness of a system.

4.3 Chapter Summary

The goal of Information Retrieval is to fulfill the information need of a user on a collection of document by identifying those documents that are relevant to the user. Since an information need is an inherently vague concept, Information Retrieval systems are based on a retrieval model that describes how informations needs can be encoded as queries and how a relevance assessment for a query and a document representation can be derived.

Digital Libraries are Information Retrieval systems that store digital documents or metadata. This work applies MTCC to a system family of three Digital Libraries for scientific information, primarily bibliographic metadata.

Two types of Information Retrieval evaluation can be distinguished, system-centric evaluation and user-centric evaluation. In a system-centric evaluation, the quality of an Information Retrieval system is assessed based on the results returned for isolated queries derived from a number of standardized information needs. The degree to which the returned documents fulfill the information need for the query is established based on relevance assessments by domain experts. Metrics like MAP, Recall and Precision describe the effectiveness of an Information Retrieval system. A user-centered evaluation of Information Retrieval does not only consider individual queries but treats Information Retrieval as an interactive process. In order to conduct a user-centered evaluation, the usability of a IR system and therefore its user interface must be taken into account.

MTCC models abstract from the user interface of a SUT, MTCC therefore does not support user-centric evaluations of Information Retrieval systems. MTCC does support the verification of Recall and Precision values for queries defined by domain experts for a collection with existing relevance assessments.

5 The MTCC Approach

This chapter gives a detailed overview of the test construction process with MTCC, the roles involved in the process and the models and artifacts used.

The MTCC testing process consists of five steps: (1) the analysis and modeling of the system family under test, (2) the modeling of testees from the system family, (3) the composition of a system-specific test model, representing the available tests for a system, and finally (4) the construction of tests represented as sequences of Test Step configurations and (5) the transformation of these abstract representations into executable Test Steps.

The MTCC process distinguishes two different roles for participants in the test construction process, domain experts and domain engineers. Domain experts provide knowledge about the functional requirements of a system, domain engineers contribute implementation knowledge and modeling skills. Section 5.1 describes the phases of the MTCC test construction process and the responsibilities of the different roles.

MTCC uses a number of different models. Models address (1) different levels of specificity — for instance at the system family level and system level — and (2) different entities such as the Services of a system and the Test Steps that exercise these Services and (3) different aspects of a system, specifically its features and its dynamic behavior. Section 5.2 introduces the different types of models used in MTCC.

5.1 The MTCC Process

The testing process based on MTCC consists of three phases followed by an external test execution phase. Figure 6 illustrates these phases and their respective activities, Figure 5.1 illustrates the relation of the MTCC Phases to the phases of the system family engineering approach as presented by Czarnecki [Cza05].

In the *Domain Engineering* phase, a system family is analyzed with respect to its test relevant Services, the features of these Services and the Test Steps needed to exercise these. Domain Engineering in MTCC is comparable to Domain Analysis [CE00, Ara89, Ara94] applied to system families, it is also analogous to Software Product Line Engineering in that core assets for the testing of the system family under test are identified and implemented.

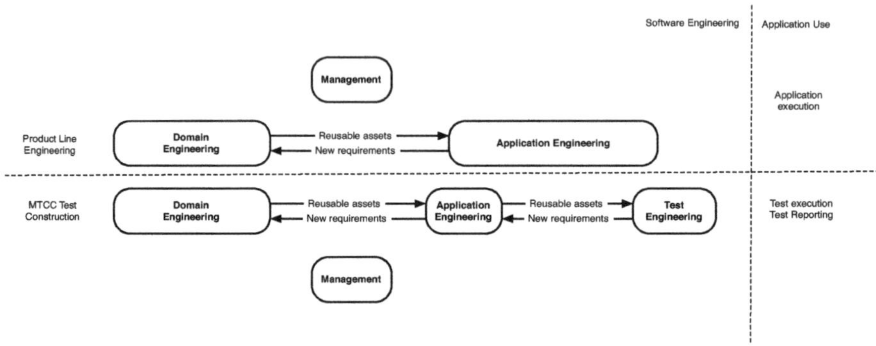

Figure 5.1: Phases of MTCC and system family engineering[Cza05]

In *Application Engineering*, the results of domain engineering are refined. Models describing the system family or domain are specialized for the representation of specific systems. The focus of *Test Engineering* is the utilization of the test editor that supports domain experts in modeling tests based on the models instantiated in the previous phases of MTCC. Test modeling with the editor is done by the selection and configuration of Test Steps, we call this Test Step instantiation. A Test Step instance represents a concrete interaction with a Service instance of the testee. The feature model of a Test Step instance is a specialization of the Test Step feature model that represents the configurations for the Test Step instance, taking into account the specific features of the Service addressed by the Test Step. The parameterization of a Test Step is expressed as the configuration of the feature model representing the Test Step instance. The use of feature models to model the inputs and expected results of a test case leads to a more fine grained use of feature models than in product configuration, where features often represent the general capabilities of a system [WS06]. In the test code generation activity, Test Step configuration serves as the base for the configuration of executable test cases.

The last phase of Testing with MTCC is not part of the core MTCC process. How the activities of this last phase, test execution and test reporting, are conducted is determined by the test runner for which tests are generated and by the general testing regime in place.

While each of the phases of the MTCC process builds on the results of the previous phase, MTCC does not assume a sequential execution of the different phases. We assume that domain engineering, Application Engineering and test engineering are executed in iterations.

MTCC can be considered a product-line approach for the construction of tests. Figure 5.2 illustates the relationship of the work on applications within a domain and the domain infrastructure in Product Line Engineering. Applied to the testing process of MTCC, three processes corresponding to the phases of the MTCC process. In test engineering, tests are constructed based on the assets implemented in domain engineering and Application Engineering. Requirements for the models and testing infrastructure flow back to Application Engineering, requirements for new classes of Services and Test Steps serve as input to domain engineering. Additions to the MTCC domain models from domain engineering are used to model new aspects of systems in Application Engineering. The core assets of the MTCC testing process are the domain models, the editor, the routines for model composition and the infrastructure for test generation. Figure 5.1 illustrates the relationship of the three MTCC phases to Application Engineering and Domain Engineering in System Family Engineering [Cza05]. The concept of System Family Engineering as used in the figure is comparable to our concept of Product Line Engineering.

The different roles involved in the overall testing process with MTCC are expected to be in constant communication. Domain experts construct tests and inform domain engineers about test-relevant, but not yet testable functionality. Domain engineers extend and maintain the models and tools used in the test construction process and oversee the generation and execution of test cases.

5.1.1 Roles and Relation to Software Development

Two roles are distinguished in the MTCC testing process, domain experts and domain engineers:

- Domain experts have knowledge about the functional requirements of the application domain or at least one system in the application domain, but do not know how systems in the domain are implemented.
- Domain engineers have knowledge about the implementation of the systems within the system family and have the necessary skills to implement and maintain the artifacts and models of the MTCC approach, but do not have sufficient knowledge about the functional requirements of the system to implement tests.

Domain engineers and domain experts cooperate in the application of the MTCC processes. MTCC assumes that the system under test is available at least as a prototype and that the execution of tests generated from MTCC models on the testee is possible.

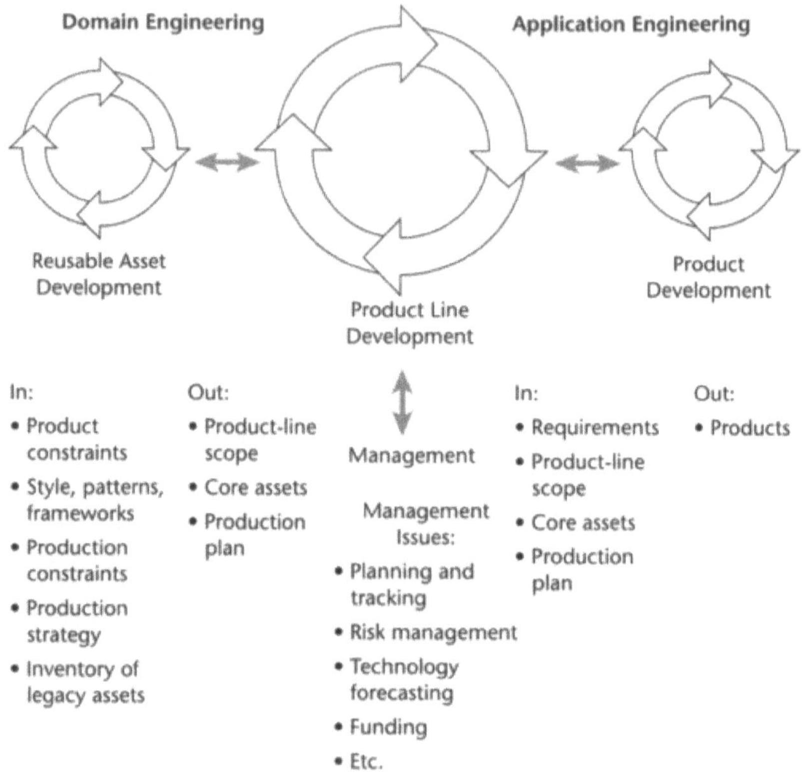

Figure 5.2: Product Line Practice as defined by the SEI Framework [CE00]

5.1.1.1 Roles in the MTCC Process

Roles in MTCC are defined by their knowledge about the functional requirements or technical implementation of the system under test. It is one purpose of MTCC to support the capture and formalization of domain experts' knowledge and operationalize it as information for the verification of systems. The capture and formalization of requirements is facilitated by the MTCC editor, a tool that allows the construction of tests for specific systems without requiring that the user possesses skills in programming or formal modeling.

The design and realization of the MTCC models as well as the implementation of the MTCC infrastructure and its maintenance are the responsibility of the domain engineer. He or she must be able to represent the test-relevant aspects of the domain identified by domain experts in the appropriate models.

We argue that the cooperation of domain experts and domain engineers is a prerequisite for the implementation of high-quality software that meets the requirements of its users. We see this notion explicitly reflected in a number of software development methods like Domain Driven Development [Eva04]. We argue that the need for cooperation applies to testing software as well.

In addition to the differentiation of roles by the respective knowledge of their holders, it is also possible to distinguish roles based on the MTCC phase they are active in. For instance, domain engineers working on the MTCC core assets need different skills than system engineers that model the Services of one particular system and maintain the Test Adapter necessary for test execution. It could be argued that domain engineers and system engineers as well as domain experts and system experts (knowing the functional requirements for one system) constitute different roles. For this work, we only use the roles of the domain engineer and the domain expert. While good reasons exist for the definition of additional roles [KRV06], we argue that for our prototype implementation of MTCC and its evaluation, such roles are not needed.

5.1.1.2 Relationship of MTCC to Product Development

MTCC defines two preconditions that the software development process whose products are to be tested must fulfill: The software development process must have incremental elements and the testable version of the SUT must be available when MTCC is used.

MTCC can only be used within the context of software development processes that have iterative or incremental elements [LB03]. The naive use of the waterfall-model [Roy87] in a strictly sequential manner as well as the classical V-Model [BR05] and all other models that do not allow for requirements discovered in testing to feed back into the design and implementation process are not well-suited for MTCC, since it is not possible for tests to influence the development process in a form of executable specification.

Another reason for the use of MTCC in iterative processes is the need for a testable system, either a prototype or a version of the final system, that supports the execution of tests and thus the decision whether the requirements defined by domain experts are meet by the system.

Since the roles in MTCC are compatible with intensive cooperation of software developers and customers in agile approaches to software development [BA04], we argue that MTCC can be used in such processes. MTCC tests can be an executable alternative to user stories in situations where MTCC models for the functionality to be implemented already exist and the implementation in the context of the system is planned next.

MTCC makes no assumption about the use of any particular method for the identification of the test-relevant features of a system and about the process used by the domain experts to determine relevant tests for these features. Methods from the field of requirements analysis [Myl05, vL01, KG03] can be used as well approaches that are explicitly designed to be used in the testing context [MSMM00].

5.1.1.3 Relationship of MTCC to the Testing Process

MTCC is an approach to acceptance testing. The MTCC approach is a supplement to other testing techniques. MTCC does not replace approaches such as developer testing or model-driven system testing, nor does it compete with approaches to SPL testing.

5.1.2 Modeling of the Domain and Individual Systems

The first phase of the MTCC approach is domain engineering. From the modeling perspective, domain engineering in MTCC is an analysis process with the purpose of identifying test-relevant functionality of the testees, followed by the formalization of the identified functionality as models.

In the Application Engineering phase, concrete instances of systems are modeled as specializations [CHE05b] of the domain models realized in the domain modeling phase. The dynamic behavior of the system is modeled as a finite state machine.

5.1.2.1 Domain Engineering

The purpose of domain engineering is the analysis of all systems within the system family for test-relevant Services and Test Steps. The results of this analysis process are formalized in models. The Domain Feature Model represents all test-relevant Services in the domain as feature models. The Domain Test Model represents the Test Steps that serve to exercise and verify these Services.

The first step of domain engineering is domain analysis [Ara89, CE00]. The subject of analysis are all features [KCH+90] of the members of the considered system family that are test-relevant. A feature is test-relevant when it is needed for the specification of acceptance tests by domain experts. In MTCC, test-relevant features are grouped into Services. A Service is a combination of features that represent some aspect of a functionality relevant for the system family. For instance, the functionality needed to start a search or a list of results for a query are both Services for the system family of Digital Libraries. Within a system family and frequently within a system, multiple different instances of a Service will exist, for example, different forms used to start a search. The variability between this different instances is expressed in the feature model of a Service.

In addition to Services, domain engineering identifies the Test Steps used to exercise a Service Each Test Step is either an action that manipulates the Service or an assertion that compares some aspect of the state of the Service with an expectation. Every Test Step is specific to one of the Services of a system family. Each Service has multiple Test Steps assigned to it. An overview of all Services and Test Steps for the system family of digital libraries considered in this work can be found in Section 8.4. The MTCC activities of Service modeling and Test Step modeling build on the results of the analysis activities and formalize them in two models, the Domain Feature Model and the Domain Test Model.

The Domain Feature Model describes each identified Service as a feature model. The structure of the feature model represents the various features of a Service the variability [CAK06, CHW98] of the feature model represents the difference that can exist between multiple instances of the same Service For the search form mentioned above, different instances can vary in the number of fields available to formulate a search as well as in the advanced search options that are supported. Each Service Domain Feature Model is expressed as a feature model with cardinalities as introduced in Section 3.2. The Domain Feature Model is a catalog for Services that exist for the considered system family. It does not describe how Services are combined to describe a system, the workflow of a system or any aspect of the dynamic behavior of the system.

The Domain Test Model represents all Test Steps that exist for the Services described in the Domain Feature Model. Every Test Step is specific to one Service it represents either an action that executes the Service and thus manipulates its state or an assertion that either verifies if some expectation about the state of the Service is true or reads some aspect of the Service state and saves it for later use. Submitting a search form and thereby starting a search is an example of an action, reading the number of results from a list of documents is an example of an assertion. Test Steps are represented as feature models that contain references to the feature model of the Service exercised, these references are unfolded in the Model Composition activity of Test Engineering and allow the inclusion of those features of the feature model of a Service instance that are relevant to test construction.

5.1.2.2 Application Engineering

In the Application Engineering phase, models for one concrete system from the considered system family are realized.

The Application Feature Model describes a system in terms of its Service instances. Each Service instance is a specialization of one of the Services described in the Domain Feature Model. It is possible and common that multiple different instances of one Service exist for a system, for instance, a simple search form for unexperienced users and a more complex search form for experts.

Details about the structure of the feature models included in the Application Feature Model and the information contained in these models can be found in Section 6.2.

The Application State Model describes the dynamic aspects of a specific system. The model is represented as a non-deterministic finite state machine that represents possible traces through the system.

The finite state machine of the Application State Model distinguishes two kinds of states: Context states and Service states. Only Context states can be reached from the initial start state of the FSM. From a Context state, only Service states can be reached. The transition from a Context state to a Service state corresponds to the selection of one Service from the set of all Services defined in a given Context. From a Service state, only Context states can be reached, the state transitions from a Service to a Context state represent the execution of a Service and the resulting change of the current Context. For example, the execution of a search Service can result in a transition to a Context representing a page that displays the result of a search or, in cases where no documents were found, a transition to a Context representing the "no-results found" page. Section 6.1 discusses the dynamic models used in detail.

Figure 5.3 illustrates the relevant concepts represented by the MTCC model at the system family and system level as well as the relations of the these concepts. A system in MTCC is represented by a number of Contexts. Every Context contains a number of of Service instances. While a Context has an unique name, it is primarily defined by the Service instances it contains. For web-applications like the Digital Libraries, a Context will typically correspond to a page of the application. A concrete example for a Context in such a system would be the page that displays the results of a search. This Context would include Service instances representing one or more lists of result documents, links to other pages of the application, and controls to manipulate a list of the results, for example by changing its sorting.

Figure 5.4 displays one page of a Digital Library that corresponds to a Context. The context includes the following seven Services: (1) Information about the

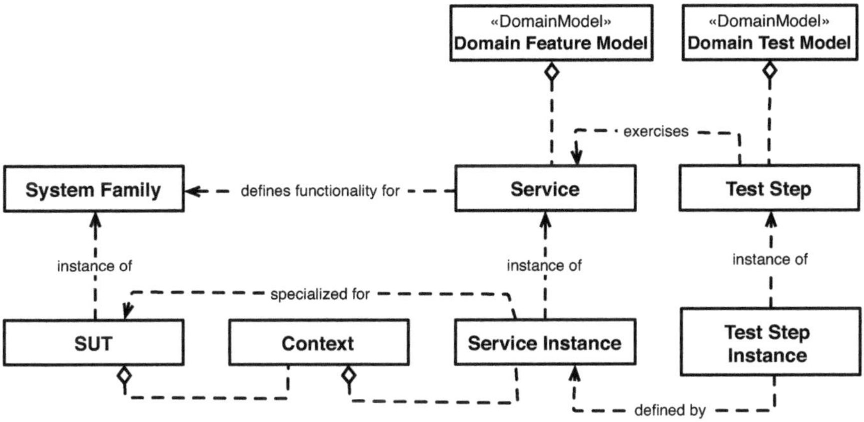

Figure 5.3: Relevant concepts of MTCC on the system and system family level

search that was sent to the digital library, (2) controls to browse the results of the search, (3) feedback about the number of documents found, (4) controls to change the sorting of a list of results, (5) facets to further restrict the results of the search, (6) controls to mark result documents for experts, and (7) a list of result documents for a search.

5.1.2.3 Implementation of the Infrastructure Code

After the Application Feature Model and the Application State Model for a testee are realized, the last activity of the Application Engineering phase is the implementation of infrastructure that is necessary for the execution of tests. Two artifacts have to be implemented in MTCC to allow the generation and execution of test cases.

Templates allow the generation of test cases for one specific test runner. The Test Adapter is a library that provides functionality needed for the execution of the previously generated test cases. The Test Adapter is specific to the system tested and to the test runner used to execute the tests.

5.1.3 Composition of Models

The MTCC editor supports the construction of tests for a member of a software family based on the information contained in the Application Feature Model, Ap-

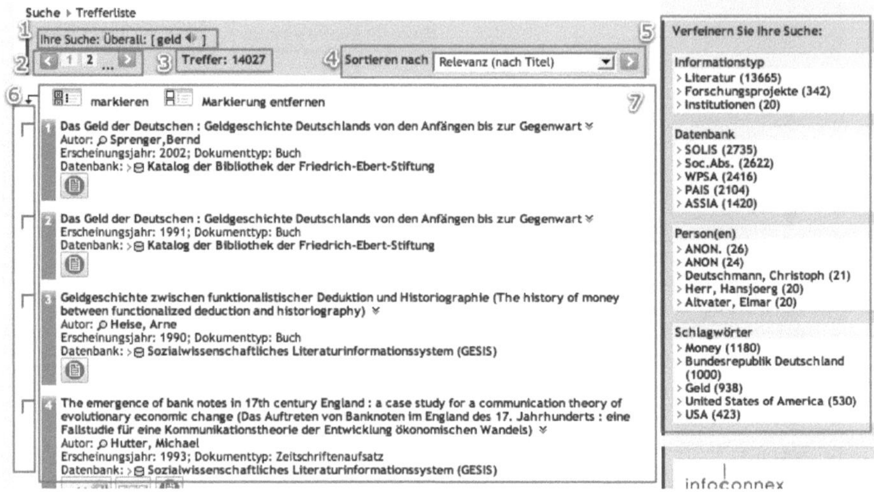

Figure 5.4: Service instances in a web page

plication State Model and Domain Test Model. The use of these models enables the editor to guide a user in the test construction process by allowing only the construction of tests that are supported by the system represented by the model.

In order to reduce the complexity of the editor, the editor does not use the Application Feature Model, Application State Model and Domain Test Model as separate models. Instead it uses an integrated model, the Application Test Model. The Application Test Model is a composition of the three initial models, it includes both information about Services and their associated Test Steps and information about the possible sequences of Test Steps that can be executed on a system.

While the Application State Model and the Application Test Model represent a system in terms of its Services, the representation used by the Application Test Model is based on the Test Steps that can be executed on a system, the configurations that are possible for the Test Steps and the possible sequences of Test Steps.

5.1.3.1 Reasons for the Composition of Models

In order to discuss the reasons for the composition of the Application Feature Model, Application State Model and Domain Test Model into the Application Test Model, we first discuss our reasons for separating these models. The decision to use three different models and only to integrate the information contained in these models in the Application Test Model immediately before the instantiation of the editor has conceptual as well as practical reasons:

- By representing Service instances in the Application Feature Model and Test Steps in the Domain Test Model, Test Steps only need to be defined once at the system family level. Service instances have to be defined at the system level.
- The definition of Test Steps at the system family level helps to avoid the redundancy that would result from the inclusion of all Test Steps for all referenced Service instances. It thus has positive influence on the maintenance of the models. For instance, in order to add a new Test Step to a Service the Test Step only needs to be added to the Domain Test Model, the Service instances exercised by the Service need not to be changed.

The information contained in the Application Feature Model and the Application State Model is best expressed with different types of models. While a feature model is by definition an excellent representation for the possible configurations of a Service instance in the Application Feature Model, a FSM is better suited to describe the possible sequences of Service invocations included as described by the Application State Model.

While the use of separate models eases the maintenance of the models, such a representation as an internal model for the editor is disadvantageous for two reasons:

- The editor needs to represent a system in terms of the Test Steps that can be executed on it, not in terms of its Services.
- The use of separate models eliminates redundancy in the models, but raises the complexity necessary to work with the models.

5.1.3.2 The Composition Process

The Application Test Model has to parts: a catalog of Test Step instances represented by feature models and a dynamic part represented by a FSM. We call the catalog of Test Step instances the structural part.

The structural part is created from the Application Feature Model by replacing every Service instance with instances of those Test Steps that are assigned to it and can be executed on it. The Test Step instances are created by checking all Test Steps defined in the Domain Test Model for compatibility with the Service instances defined in the Application Feature Model. If the Test Step and the Service instance are compatible, a Test Step instance is created by unfolding the feature references in the Test Step and is added to the catalog.

The dynamic part of the Application Test Model is created by rewriting the Application State Model in such a way that Service state is replaced by a number of Test Step states in the same way that Test Step instance replace Service instances. The resulting FSM describes all possible sequences of Test Steps that can the executed on a testee.

5.1.4 Modeling of Test Configurations

The editor supports the construction of test cases based on the dynamic and structural information in the Application Test Model.

From a technical point of view, a test is constructed by building a sequence of configured Test Step instances. The FSM in the dynamic part of the Application Test Model determines which sequences of Test Steps are available. The feature models of the structural part define how each Test Step instance can be configured.

From the users point of view, test construction is defined by two activities: The selection of Test Steps that represent a usage scenario that is to be tested and the parameterization of each individual Test Step.

Like Action Word [BBN04] based approaches to testing or FIT [MC05b] workflow tests, MTCC represents a test as a sequence of individual steps that manipulate and verify the testee. The combined Test Steps represent one possible session of a user with a system. From a technical point of view, a MTCC test is a trace through the functionality of the system. One property specific to MTCC is that the selection and combination of Test Steps is guided by the editor. A user can only construct those sequences of Test Steps that are supported by the system under test as described by the finite state machine of the Application Test Model. For example, a Test Step to submit a query would only be available if such an action would be supported by the testee in its current state.

The selection of Test Steps determines the functionality of a testee that will be exercised by the test. The parameterization of the individual Test Steps determines exactly how the Service instances of the testee are exercised and verified. The possible parameterization or configuration of each Test Step instance is defined by feature models in the structural part of the Application Test Model. The editor represents the specialization steps that can be applied to the feature model in terms of the elements of a graphical user interface. As it is the case with the selection of Test Step instances, the editor uses information from the Application Test Model to guide the user in test construction. A test in MTCC is a sequence of configured

Test Step instances for one system of the system family. We call such a test a Test Configuration. An in-depth discussion of the MTCC editor is given in Section 7.1, details on the specialization of feature models for the purpose of the configuration of Test Step instances are given in Section 7.2.

5.1.4.1 Representation of Tests in the Editor

The editor displays the sequence of Test Step instances that define the behavior of a test as a list on the left part of the GUI. The Test Step instance that is currently being configured is displayed in the right part of the editor. The graphical user interface that represents each Test Step is dynamically built by the editor based on both the structure of the feature model of the Test Step instance and the semantics of individual nodes in the feature model. Figure 5.5 shows the MTCC editor. The Test Steps for the current test are shown in the left part of the GUI (1), the right part of GUI (2) displays a representation of the current Test Step.

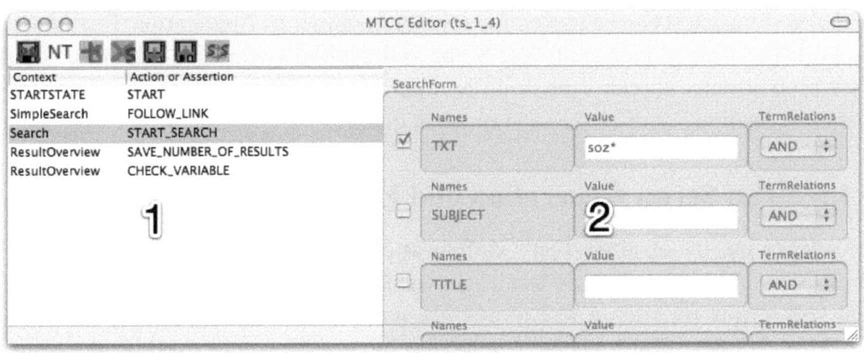

Figure 5.5: Representation of the available and current teststeps in the MTCC editor

In a breadth-first traversal of the feature model, each node is checked against a number of GuiBuilder objects. Each GuiBuilder object is responsible for the display of specific nodes or subtrees of a feature model. When a GuiBuilder object is responsible for the display of a subtree of a feature diagram, it either builds a complete representation of GUI elements for the feature model or a partial representation and delegates the rendition of some subtree to other GuiBuilder object, similar to the application of the *Chain of Responsibility Pattern* [GHJV95].

The configuration of a Test Step instance done via the editor is represented by configuration nodes in the feature model. Configuration nodes are attached to the nodes of the feature model that are configured by the user. For instance, when an optional feature is de-selected by clicking the checkbox that represents that feature in the GUI, a configuration node representing refinement of the feature cardinality from [0,1] to [0,0] is attached to the feature node.

The representation of the specialization steps used to configure a Test Step instance as nodes of the feature model instead of directly applying them to the model has advantages for the representation of feature models in a GUI. Since the specialization steps used during configuration do not directly manipulate the feature model but are represented as a separate object, it is easy to undo and change the configuration of a feature model.

5.1.4.2 Reuse of Tests

One purpose of MTCC is the reuse of tests, more precisely of Test Configurations for multiple systems. Since a Test Configuration is constructed from the Test Step instances for one particular system as represented by the Application Test Model of that system, a test can be reused for another system if its Application Test Model supports the same sequence of Test Steps as the initial system and each individual Test Step instance supports the same configuration as the corresponding Test Step instance. Section 7.3 discusses test reuse in detail.

5.1.5 Test Script Generation and Execution

In order to use the abstract Test Configuration for testing, MTCC employs code generation. MTCC generates test cases for COTS or test runner. It is assumed that test scripts for such a test runner are compatible with the structure of xUnit [Mes07] tests. The test cases generated by MTCC do not exercise the testee directly but use a manually implemented wrapper library, the Test Adapter. The Test Adapter provides implementations for each Test Step instance in the system family for a specific pair of system and test runner.

With the use of code generation and the Test Adapter, MTCC facilitates the decoupling of the semantics of the functionality to be tested and its implementation. Since MTCC does not assume the use of any particular test runner, the integration of MTCC in an existing testing process is also supported.

5.1.5.1 Generation of Test Code

Compared to other applications of code generation, the code generation approach in MTCC is relatively simple. Since test cases neither include complex data structures nor complicated control flow, a template-based approach to code generation is sufficient.

Every Test Configurationobject is represented by one test case. The test case is a sequence of method calls for methods defined in the Test Adapter. Each method corresponds to one configured Test Step instance in the Test Configuration.

The generation of test case code is done in two phases, in the first phase, the feature model of each configured Test Step instance is transformed into a simplified representation as a set of key-value pairs. This presentation is called the parameter object. In the second phase of the code generation process, method calls for each Test Step instance in the test are looked up from the Test Adapter library and code for the invocation of the method with the parameter object as its parameters are generated. Section 7.4 discusses the MTCC approach to test code generation. The execution of the generated tests as well the logging and reporting of test results are not part of the MTCC process.

5.2 Overview of the MTCC Models

The models used in the MTCC are directly related to the phases of the MTCC testing process. In Domain Engineering, Test Steps and Services relevant to the system family as a whole are expressed in the Domain Test Model and the Domain Feature Model respectively. Application Engineering derives the specific Services instances for one system by specializing the Domain Feature Model into the Application Feature Model. The Application Test Model describes the possible sequences of Service invocations for a system. In test engineering, the Domain Test Model, the Application Feature Model and the Application Test Model are merged into one composed model that describes a test in terms of the Test Step instances that can be executed on it. The editor uses the Application Test Model to support the construction of Test Configurations, sequences of configured Test Step instances that are an abstract description of a test.

Figure 5.6 illustrates the relations of the models used in the MTCC process and their associated phase of the MTCC testing process. Chapter 6 discusses the information contained in the models and their implementation in detail.

5.2.1 Domain Level Models

The MTCC domain level models represent the test-relevant aspects of all members in a system family modeled by MTCC. MTCC uses two models for the representation of test relevant models, the Domain Feature Model and the Domain Test

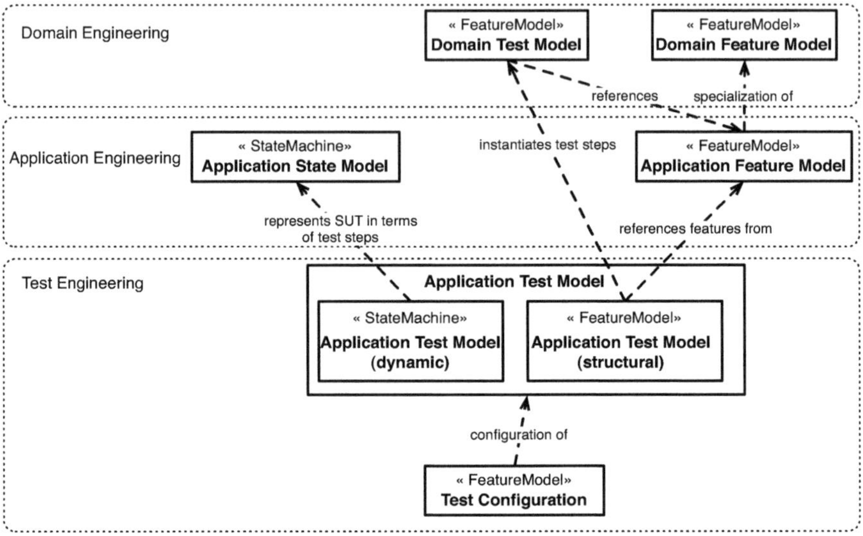

Figure 5.6: Models used in MTCC and their relationships

Model.

The Domain Feature Model represents the commonalities and variability for each type of Service present in the system family. It is usually implemented as a feature model with cardinalities[1]. The Service instances of the specific members of the system family are specializations of the Services defined in the Domain Feature Model. By deriving all Service instances from a common model, the comparability of individual Service instances is ensured. This is an important factor when the features provided by one Service instance are compared with those of another Service instance in order to check if a Test Configuration can be transfered from the first Service to the second.

The Domain Test Model is a catalog of Test Steps. Each Test Step is specific to one class of Service Test Steps do not include information about the structure or variability of the Services. Instead, they contain references to the aspects of a Service instance that is relevant for the execution of the Test Step. MTCC models at the domain level do not have information about the dynamic behavior of the testee.

[1]For tooling reasons, we implemented the Domain Feature Model used for the evaluation as a class library that instantiates Application Feature Model models directly.

5.2.2 System-Level Models

The MTCC System-Level models created in Application Engineering describe one system in terms of its Service instances and the possible interactions with these Service instances.

The Application Feature Model represents Service instances of one particular system within a system family. Service instances are feature models that are specializations of the Service classes expressed in the Domain Feature Model. Multiple instances of a Service can exist within one Application Feature Model. The relationship of a Service and a Service instance is analogous to the relationship of a class and an object in object oriented approaches.

To give an example, the representation of a search form Service in the Domain Feature Model would describe the kinds of fields available for search and the options to influence the search that can exist for a system family. The Service instance in the Application Feature Model describes one concrete search form for one system.

The feature models of Service instances are not fully specialized. The remaining variability for representing the parameters that can be used for the invocation of a Service The Application State Model describes all possible interactions between a user and the SUT as possible sequences of Service invocations. The Application State Model is implemented as a finite state machine. The Test Steps defined in the Domain Test Model are not refined on the system level.

5.2.3 Test-Level Models

MTCC uses two models at the test engineering level: The Application Test Model serves as input to the editor, the Test Configuration represents tests on a system.

The Application Test Model is a composition of the Application State Model, the Application Feature Model, and the Domain Test Model. It does not include any information or refinements not already included in one of its input models. Instead, the Application Test Model is an alternative representation of the information included in the other models that is explicitly designed for the purpose of facilitating test construction using the MTCC Editor.

As can be seen from Figure 5.6, the Application Test Model consists of a structural part and a dynamic part. The dynamic part is derived from the Application State Model. It is a finite state machine that describes all possible sequences of Test Steps that can be applied to a SUT.

The structural part is instantiated by unfolding the references from the Test Steps defined in the Domain Test Model to the Service instances defined in the

Application Feature Model. This unfolding of references creates instances of Test Steps whose features are a combination of the feature model of the Test Step and the feature model of a Services instance.

We use the name Application Test Model for both the dynamic and structural aspects of the model to emphasize that the Application Test Model and all its parts are created in one process and only for the purpose of test construction. The composition of the Application Test Model is a fully automated process that occurs immediately before test construction in the editor. The Application Test Model is a transient model, it is only created for the purpose of test construction and is never persisted.

Test Configuration Models are abstract descriptions of tests for one system. Conceptually, these models represent a test as a sequence of configured Test Step instances. The feature models that represent the Test Step instances are fully specialized.

5.2.4 Relationship of MTCC Models to the MDA

Since MTCC uses models of the different levels and steps of refinement and thus exhibits parallels to concepts used in the MDA, the question arises how the hierarchy of CIM, PIM and different levels of PSM [MSU04] relate to the domain- and system-level models of MTCC.

One significant difference is that the different models used in the MDA approach represent different levels of abstraction with respect to an implementation. The MTCC models at any level, in contrast, are agnostic to the final implementation of the tests. In MTCC, different levels of modeling differ not in their details with regards to the implementation, but in the scope of the modeled entity like system family, system and test.

5.3 Chapter Summary

The MTCC process is in many aspects similar to Software Product Line Engineering. Where the Software Product Line Process is defined by two parallel activities, Domain Engineering and Application Engineering, MTCC consists of three phases, Domain Engineering, Application Engineering and Test Engineering. The focus of Domain Engineering is the analysis and modeling of the test-relevant features of the system family and the implementation and maintenance of the MTCC Core Assets. The Application Engineering phase considers a specific SUT. The models created in Domain Engineering are specialized to represent the features available for the SUT and the artifacts needed for text execution are implemented. The products of the MTCC approach are automated acceptance Tests. These tests are created in the Test Engineering phase. In Test Engineering, domain experts use the MTCC editor to create an abstract test representation. These abstract representations are then used to create executable tests.

The models used in MTCC can be distinguished by the phase in which they are used and by the aspect of the tested system that is represented. The features of a SUT, the tests used to exercise these features and parts of the dynamic behavior of the SUT are presented by feature models and finite state machines. Domain-level models represent the Services and Test Steps that are relevant for all members of the system family. System-level models represent the behavior of a system and the instances of Services that are available for the system. Test-level models combine the information included in the system and domain-level models and represent tests.

MTCC partitions a system into basic units of functionality called Services. Services are grouped in Contexts. Contexts and the Services defined in Contexts define the testable interface of a testee. The actions that are available to exercise and verify the SUT are represented as Test Steps.

6 Models for System and Test Representation

This chapter discusses the models used in MTCC in detail. While the proceeding chapter focused on the utilization of models in the MTCC testing process and the different models used at different levels of abstraction, this chapter focuses on the implementation of the models.

Section 6.1 describes the Application State Model, the model used by MTCC to represent the behavior of a system. MTCC employes a finite state machine for the Application State Model, specifically a UML Protocol State Machine [RJB04], to represent the possible interaction of a user or test with the system under test.

In section 6.2, we discuss the use of feature models [CHE05c, KC05] in MTCC to represent both the Services of systems under test and the Test Steps that are applied to these systems. The Services of a system that are to be tested are represented by the Domain Feature Model and the Application Feature Model. A Service in MTCC is the feature model representation of some functionality recurring for multiple systems in the system family under test and testable in isolation. The representation of a Service instance at the system level includes details about the functionality provided by this particular Service especially as they concern the interface of the Service

The Domain Test Model is a domain-level structural model that describes the Test Steps that can be applied to the Service instances defined in the Application Feature Model.

Section 6.3 discusses the composition process that merges the Application Feature Model, Domain Test Model, and Application State Model into the Application Test Model. The Application Test Model is a transient model that represents the available tests for a member of a system family. The model is optimized to support test case construction using the editor. The Application Test Model includes a dynamic part derived from the combination of the Application State Model and the Domain Test Model as well as an dynamic part that is created by unfolding the references from the Test Steps in the Domain Test Model to the Services instances in the Application Feature Model.

6.1 Models for the Behavior of a Testee

The Application State Model and the dynamic part of the Application Test Model describe the behavior of the SUT. Both models are represented as finite state machines where states represent the selection and execution of Services or Test Steps respectively. The Application State Model and the Application Test Model are related as follows:

- The Application State Model is a System-level model. It describes a system as a number of Contexts, each of which contains a number of Services. The transitions between Context and Services states represents the selection and execution of Service The Application State Model is modeled during the Application Engineering phase. It does not include any information about the actions or assertions available to test the Service
- The dynamic part of the Application Test Model is a test-level model that describes all sequences of the Test Steps that are available for the manipulation and verification of a testee. This model is a combination of information from the Application Test Model and the Test Steps definitions from the Domain Test Model.

MTCC does not define any models of the behavior of systems at the system family level. We argue that the differences of systems within a system family regarding their supported interactions with Services, are so significant that domain-level behavior models are not useful in practice.

The reason for the use of both the Application State Model and the dynamic part of the Application Test Model to represent the behavior of a testee is the intent to separate the representation of a system and tests for the system and thus ease the creation and maintenance of the models. By limiting the Application State Model to the representation of models, redundancies are avoided. The Application State Model is thus the canonical model for the behavior of a system, while the dynamic part of the Application Test Model is a representation derived from the Application State Model for the purpose of Test Step selection during test construction.

As discussed in the overview of the MTCC models in Chapter 5, the dynamic and the structural part of Application Test Model are transient models created in a fully automated process immediately before the use of the editor for test case construction. Since both the dynamic and the structural part of the Application Test Model serve the same purpose and are created in the same process, we treat the Application Test Model as one model. From a technical and conceptional point of view, the dynamic and structural parts of the Application Test Model are different models, this section discusses only the dynamic part, an examination of the structural part follows in Section 6.2.3.

6.1.1 Structure of the MTCC Behavior Models

As already discussed in Section 5.2, the models used in MTCC for the representation of the behavior are finite state machines. We use the UML2 Protocol State Machines to illustrate the structure of FSMs in this work where appropriate, but do not assume that the exact semantics of Protocol State Machines are implemented in MTCC.

From a conceptional point of view, MTCC differentiates between three kinds of states: Context states, Service instance states and Test Step instance states. The Application State Model contains Context and Service instance states. The dynamic part of the Application Test Model replaces each of the Service instance states from the Application State Model with one or more Test Step instance states.

Figure 6.1 illustrates the relations of the different concepts relevant for the dynamic models used in MTCC. A system is represented as an aggregation of Contexts. Each Context is in turn an aggregation of Service instances or Test Step instances. In the FSM, Contexts are presented as states from which the aggregated Service Test Step instances can be reached. In the following, we discuss the structure of the FSM that represents a testee in the Application State Model as well as in the dynamic part of the Application Test Model.

6.1.2 Concepts and Structure of the Application State Model

The Application State Model represents all possible interactions with a testee as a sequence of Service invocations. Each invoked Service instance is selected from all Service instances defined for a Context. A testee consists of a number of Contexts, from which most Service instances can be reached. Each Service instance in the Application State Model can be reached from exactly one Context. The invocation of a Service instance leads to the transition of the system under test into a new Context. The new Context can either be the initial Context, from which the invoked Service instance was selected or a new Context. A Service instance can have multiple different Contexts that can be reached as the consequence of an invocation. For instance, one Context that represents the state of the system after a successful Service execution and another that represents the state of the testee after a failure occurred.

Figure 6.2 displays an example Protocol State Machine for a simple system that

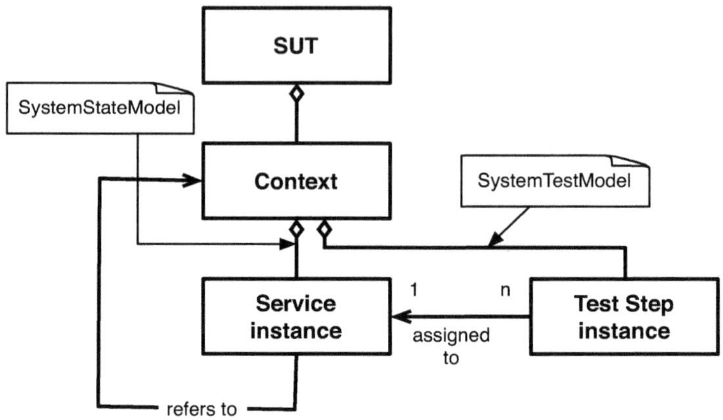

Figure 6.1: Relationship of the concepts in the dynamic MTCC models

consists of three Contexts and three Service instances. Three types of states and three types of transitions are used in the FSM. The Context and Service states represent Context and Services as described above. The transitions are named after the event that triggers the state transitions. The three types of transitions that are used in the Application State Model in addition to the transition from the start state are select Test Step, exercise service, and end test:

- The transitions triggered by the select Test Step event cause the change of the current state of the FSM from a Context state to a Service state. These transactions represent the selection of one Service instance, for example, a search form or an link, but not yet the execution of this Service instance.

- The transition exercise service changes the current state of the FSM from a Service state to a Context state. This transition represents the execution of a Service

  Starting from a Service state, multiple different Context states can be reached by an exercise service transition, the FSM is thus non-deterministic. Contexts reached by different exercise service transitions represents different outcomes of Service execution. The different outcomes include, but are not limited to, the successful and the failed execution of a Service

  In the example given in Figure 6.2, the successful execution of each of the three Services leads to a transition to either state Context 1 or Context 2. A failed Service execution leads to the transition to the state Failure.

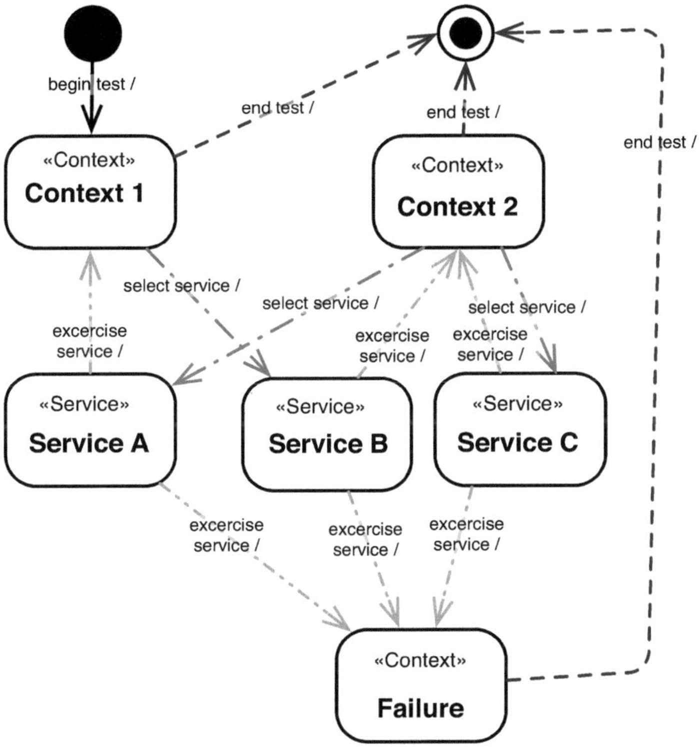

Figure 6.2: Representation of the Application State Model as a Protocol State Machine

- The end-state of the FSM is reached by one of the end test transactions. An end test transaction is available from every Context state in the FSM, it represents the end of a testing session.

Only some of the Context states of the Application State Model can be reached from the start state of the FSM. The states that are reachable from the start state represent those Contexts of a system in which a testing session can start, for instance, the home page of a Digital Library. In Figure 6.2, the only Context state that can be reached from the start state is Context 1. All tests must thus start with one of the Service instances available from this state.

The Application State Model is only a limited representation of the behavior of a system. It only includes enough information to represent the available sequences of Service invocations for a system. Information about the Service instances themselves beyond their possible successor Contexts are not part of the model. Particularly, the Application State Model does not model the internal state of Service instances.

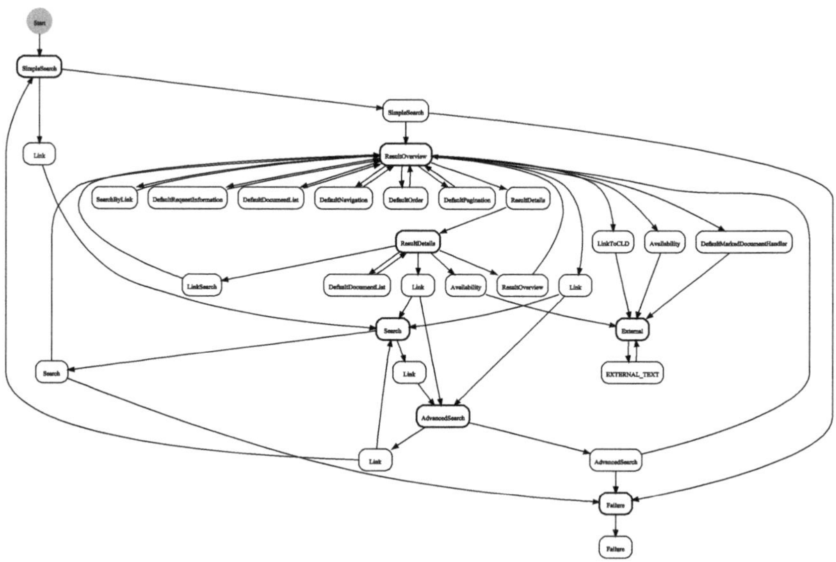

Figure 6.3: Excerpt from the Application State Model of the Sowiport System

Figure 6.3 display an excerpt from the Application State Model of the Sowiport system, rendered by GraphViz[1] exporter. This representation displays Context Nodes with thick borders.

6.1.3 Test-Level Behavioral Models

While the Application State Model represents possible sequences of Service invocations for a system, the dynamic part of the Application Test Model describes available sequences of Test Step instances. The dynamic part of the Application Test Model is created by the composition process discussed in Section 6.3. Each Service instance is replaced by a set of Test Step instances. The Test Steps defined

[1]http://www.graphviz.org

in the Domain Test Model are tested for compatibility with each of the Service instances defined in the Application Feature Model. If a Test Step is compatible with a Service instance, the Test Step is instantiated, meaning that the references of this Test Step to the Service instance are unfolded and the referenced feature nodes are replace the references in the feature model [CHE05c].

Figure 6.4 gives an example for the FSM of the dynamic part of the Application Test Model. The system in the example consists of three Contexts and six Test Step instances.

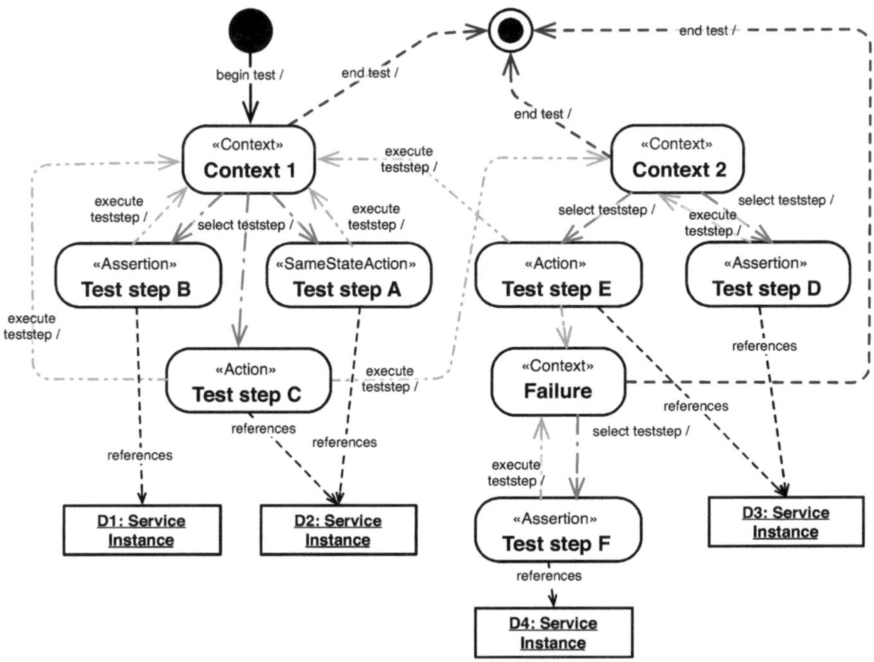

Figure 6.4: Protocol State Machine of all possible sequences of Test Step instances in the Application Test Model

The transition types of the dynamic part of the Application Test Model are similar to those in the Application State Model. Since Service instances in the FSM are replaced by Test Step instances, the names of the transitions change from select service to select Test Step and from execute service to execute Test Step. MTCC uses three types of Test Step instances in the dynamic part of the Application Test Model: Actions, SameStateActions, and Assertions.

- A Test Step instance of the type Action interacts with the testee and manipulates its internal state. The execution of this Test Step will generally result in the transition to a different Context than the one from which the Test Step instance was selected. Test Steps of the type action represent actions of a user to change the currently active page of a web application. A Test Step instance that sends a query to a Digital Library for instance can change the current Context to the result overview.
- A SameStateAction represents an interaction between a user and the testee that cannot result in a change of the current Context. After the execution of a SameStateAction, the system always returns to the initial Context. An example for a SameStateAction is the selection of filter values in a search form. While this interaction changes the state of the testee, it does not result in a change of the currently available Services.
- The Test Step Assertion represents an action that does not manipulate the state of the testee but either compares the state of the testee with an expectation or saves some aspect of the state of the testee in a variable. The execution of an assertion always returns the FSM to the Context state that was active before the execution of the test.

Similar to the Application State Model, the dynamic part of the Application Test Model only describes possible sequences of Test Step instances.

The testee and the execution test runner contain internal state information. For example, the test runner used in MTCC stores aspects of the externally visible state of the test runner in variables that can be used to compare the actual and expected behavior and state of a system. Such state information is not represented in the dynamic part of the Application Test Model. MTCC is a Black Box [Bei95] approach to testing. It does not interact with or models the internal state of a testee, thus the FSMs used in the Application State Model and the dynamic part of the Application Test Model do not represent this information.

6.2 Feature Models for System and Test Representation

The purpose of the structural MTCC models is the representation of the test-relevant aspects of a system and the Test Steps that exercise these aspects. The structural models are expressed as feature models with cardinalities. The domain-level Domain Feature Model and the derived, system-level Application Feature

Model represent the testable functionality of a system as the number of Services or Service instances. The Domain Test Model contains models for all Test Steps that can be executed on these Services.

6.2.1 Feature Modeling in MTCC

In MTCC, feature models are used to describe the Services that provide the test-relevant functionality of a system. MTCC takes an external view on the Services that focus on the external interfaces provided by the Service as it is seen by a user or an automated test. The feature models therefore represent both high-level features like the availability of specific operators in search and low-level details about the different filter values that can be used to constrain the results of a search.

Test Steps are represented as feature models that reference the feature models of Services. Test Steps themselves are system family level models that are not specialized for any system or Service of a system. System-specific information is included in feature models by unfolding the references to the Service instances in the Application Feature Model.

The feature models used in MTCC are a variant of feature models with cardinalities as introduced by Czarnecki [CHE05c]. This kind of feature model is an extension of simpler forms of feature models [CE00] which in turn are based on feature models defined in the FODA [KCH+90] context.

Compared to FODA feature models, feature models with cardinalities have the following extensions [CKK06] relevant in the context of MTCC:

- Feature groups are treated as independent nodes in a feature model. Feature groups support the selection of a subset of all children of a feature group.

- Attributes provide a way to save textual or numeric information in feature models without being limited to predefined values. One application of attributes is to assign names to cloned features or to assign numbers from a potentially unlimited domain. MTCC uses attributes to save test entries by users in feature models and thus allows, for example, the pre-filling of search forms in Test Step instances that submit a query. Attributes in MTCC are untyped and can save a textual value.

- References support the linking and inclusion of subtrees from one feature model in another. References therefore help to avoid redundancies. Where one part of a feature model would have to be replicated without references, with references it can simply be referenced from multiple nodes within the feature model. In MTCC, attributes are used by Test Steps to refer to the features of Service instances that are relevant for the parameterization of the Test Steps.

- Cloning supports the repeated replication of a subtree of the feature model. The clones of a feature are not different from other features in a feature model and can be configured independently from each other. MTCC does not use cloning in test modeling: Cloning is not implemented in the MTCC prototype used for evaluation in the context of this work, but cloning would be useful when features are replicated during the derivation of a Application Feature Model from the Domain Feature Model.

In addition to the concepts above, MTCC extends feature models with cardinalities by the addition of configuration nodes. These nodes represent specialization steps that are appended to a feature node when the node is specialized. Configuration nodes are discussed in detail in Section 7.2.

The particular implementation of feature models with cardinalities that we implemented for this work has a number of limitations. One limitation concerns the possible cardinalities for feature nodes. Cardinalities other than [1,1] and [0,1] are currently not supported. References are implemented as attributes that contain the path of the referenced nodes.

On the implementation level, MTCC uses XML to represent feature models. XML is used for the serialization of feature models. A XML object model is used as the superclass for the class hierarchy of the MTCC feature nodes. Figure 6.5 displays the classes used by MTCC to represent a feature model and the attributes of these classes. Listing 6.1 gives an example for the representation of a feature model as XML.

```
<Feature max="1" name="Variable" min="1">
  <FeatureGroup max="1" min="1">
    <Attribute value="Wert␣1" />
    <Attribute value="Wert␣2" />
  </FeatureGroup>
</Feature>
```

Listing 6.1: XML representation of a feature model

In contrast to other approaches [CPRS04] that use XML to implement feature models, MTCC does not use schema languages to describe the possible structure of a feature model and the specialization steps that can be applied to it. Instead, we implement methods for the members of our class hierarchy that allow the individual nodes to report their available configurations.

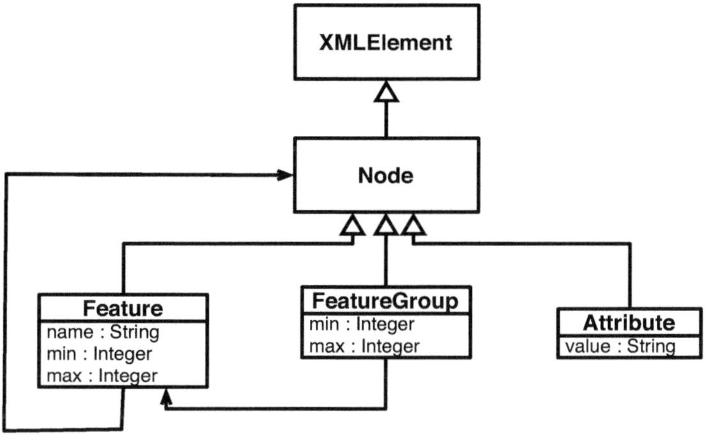

Figure 6.5: Hierarchy of the MTCC feature classes

Since the methods that implement the application of specialization steps to the MTCC models include checks that ensure that only valid transformations can occur, we argue that schema validation is not needed.

In relation to other notations used for the representation of XML feature models, like the textual notation introduced by Deursen [vDK02], we argue that while XML is more verbose than specialized notations, the tool support available for XML together with the fact that the XML models are not intended to be edited by humans makes a XML-based approach better suited in the context of MTCC.

The utilization of feature models in MTCC is invisible to the domain experts that are involved in test construction. MTCC does not make any assumptions about the tools used by the domain engineers. While MTCC does not include an editor for feature models like the one presented by Antkiewicz [AC04] that would allow the direct manipulation of feature models, the utilization of such an editor to support the realization of the Domain Feature Model and the derivation specialized Application Feature Model should be beneficial.

6.2.2 Modeling of Test-Relevant Services

By using feature models, an initial general description of the test-relevant aspects at the domain-level can be refined into models for system description and finally into models that facilitate the construction of tests for a system. On the system-family level, the overall features of Services are represented as Services in the Domain Feature Model. Following an analysis of a system from the system family, Service instances are realized in the Application Feature Model by applying specialization steps to the feature models of Services. The Test Step instances defined in the structural part of the Application Test Model duplicate those subtrees of the Application Feature Model Service instance feature models that are relevant for the Test Step. In the test modeling, the feature models that represent these aspects of Service instances as well as the feature models that define the interface of a Test Step are configured to a concrete Test Step without any remaining variability. A sequence of such configured Test Step instances, a Test Configuration, represents a test.

With that said, the modeling of tests with MTCC is partly an application of the staged configuration process [KC05, CHE05b] for system construction to the domain of test construction.

In addition to the representation of the test-relevant features of a model, MTCC also uses feature models to represent the overall structure of a testee. A SUT in MTCC is described as a hierarchy: The system contains of a number of Contexts, which in turn consist of a number of Services. Figure 6.6 illustrates this hierarchy and the classes and attributes at the different levels of this hierarchy as a UML class diagram. Figure 6.7 gives an example for the overall structure of a system as a feature model. Each Service instance is represented by a feature model that includes both attributes like the unique name and the type of the instance of well as the test-relevant features that determine the testable interface of the functionality represented by the Service instance. The information about the testable features is defined in the sub-nodes of the interface node of the feature model. The features outside the interface node describe the Service instance and are not variable after Service instance instantiation.

As Figure 6.6 displays, SUTs, Contexts and Services each have an unique name. MTCC uses this name to differentiate and to reference different Contexts and different instances of one Service within a Context. The type attribute of a Service instance informs about the Service that this instance was derived from. The type of a Service instance determines whether a Test Step defined in the Domain Test Model can be applied to the Service each Test Step is specific to one type of Service The MTCC models include documentation for each type of Service that is defined for a system family.

The structure of the interface-nodes of the feature model is specific for each type of Service The different Services for the system family of Digital Libraries considered for this work are discussed in detail in Chapter 8.

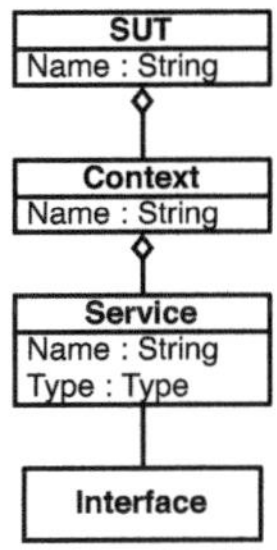

Figure 6.6: Entities used in MTCC to represent the structure of a system represented as a class hierarchy

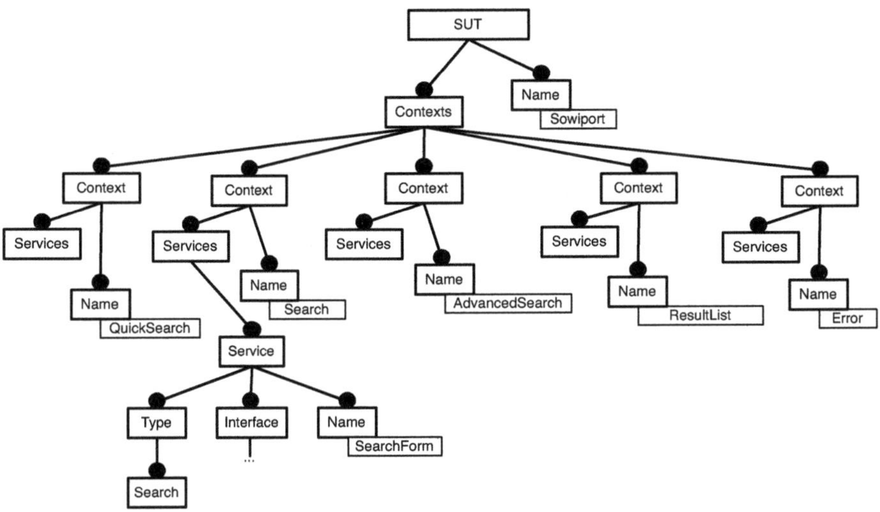

Figure 6.7: Example feature model for a system in MTCC

Figure 6.8 displays a part of the feature model of a Service instance of the SEARCH_FORM Service Compared to the feature model of the SEARCH_FORM Service defined in the Domain Feature Model, this feature model is specialized to a certain degree. For instance, except for one feature, all children of FieldType and FieldCombination feature groups are de-selected.

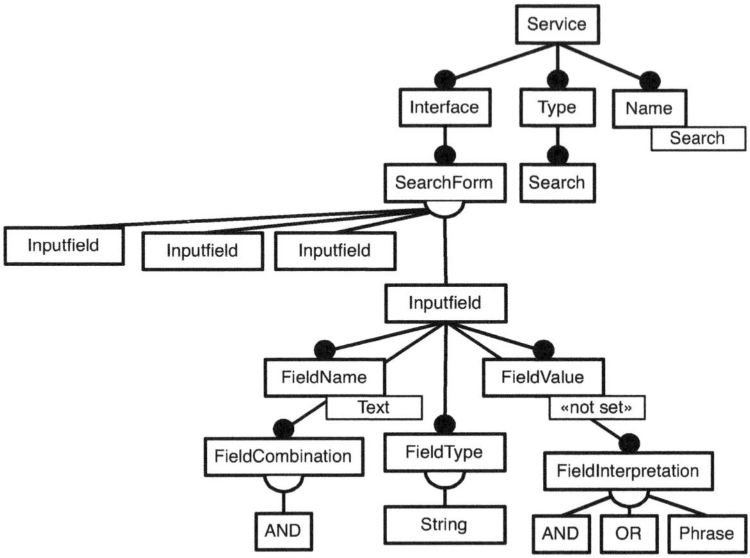

Figure 6.8: Part of the feature model of a Service instance

6.2.3 Representation of Test Steps

Test Steps represent actions that can be used to manipulate the state of a testee as well as assertions that either compare an aspect of the state of the testee with some expected value or save such an aspect of the state in a variable for later use. All Test Steps for a particular system family are defined in the Domain Test Model. Every Test Step is specific to one type of Service During composition, all Test Steps are checked for compatibility with all Service instances that are of the type required by the Test Step. Details on the composition process are given in Section 6.3.

In addition to the type of Service instance, Test Steps can define a number of additional criteria called checks in MTCC, that can be used to test whether a Test Step can be applied to a particular Service instance.

If a Test Step and a Service instance are compatible, a number of zero or more references that each step defines for parts of the feature model of a Service instance are unfolded. The referenced subtrees are thus copied into the feature model of the Test Step. We call this unfolding of references the instantiation of the Test Step. The resulting model is called the Test Step instance. The structural part of the

Application Test Model contains all Test Steps instances that can be instantiated for a given system. From the perspective of automated testing, Test Steps correspond to action words [GF00] that represent particular functionality that can be executed on a Service Test Step instances thus can be seen as specialized action words that support the execution of one particular Service and which define an interface that represents the testable features of this Service instance. Test Steps in MTCC contain the following information:

- A unique name that is used in the editor to describe the functionality and executed by the Test Step in the editor. When the Test Step is instantiated, this unique name together with the name of its exercised Service instance defines the Test Step instance.
- A number of checks that are executed against a Service instance and on which a decision on the compatibility of the Test Step and the Service instance can be determined. The most basic of these checks that are defined by all Test Steps in MTCC is the check for the type of a Service instance.
- Features that define the interface of the Test Step. This includes both feature models that represent choices specific to the Test Step, such as the choice between different variables that can be used for the comparison of a value or an attribute that allows the specification of a value, and references that include parts of the feature model of a Service instance into the feature model and hence the interface definition of a Test Step instance.

From a purely conceptional point of view, the use of feature models for the representation of Test Steps is not necessary since Test Steps are not specialized but rather include specialized, system-level information. We argue that the use of feature models for the representation of Test Steps is justified by the advantages that result from the use of feature models as a common metamodel for both Services and Test Steps. The use of a common metamodel allows the reuse of existing tooling and usually creates less complexity than would result from the utilization of different metamodels. We further argue that feature models are well suited to represent the parameterization of Test Steps. Figure 6.9 illustrates the different concepts included in a Test Step as a UML class diagram.

- Checks are used to test a Service instance regarding its compatibility with the Test Step. Test Steps in MTCC support three different classes of checks:
  - An instance of the class TypeCheck is used by every Test Step. This check tests whether the type of a Service instance is equal to the Step-Type attribute of the Test Step. Only when that condition is fulfilled,

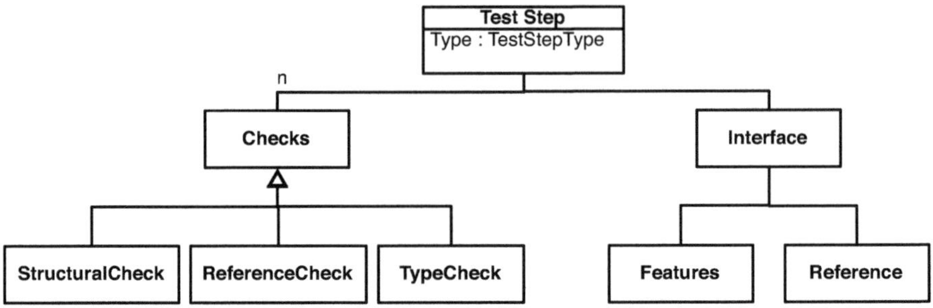

Figure 6.9: Class diagram of the relevant concepts of a MTCC Test Step

are the types of Test Step and Service instance compatible and any remaining checks of the Test Step need to be tested.

- Instances of the class ReferenceCheck ensure that a reference to the feature model of a Service instance can be unfolded. When the referenced features are not present in the Service instance, the check fails and the Test Step cannot be instantiated for this Service instance. In contrast to the references defined in the interface part of the Test Step feature model, no copy of the referenced feature is made when the references defined in ReferenceCheck are unfolded.

  In the implementation of MTCC, the references used as ReferenceChecks are expressed as XPath statements.

- Instances of the class StructuralCheck support the test of Test Steps against Service instances based on the structure of the feature model of the Service instance as defined by a schema language. As is the case with the ReferenceChecks discussed above, StructuralChecks in the current implementation address the underlying XML representation of the MTCC feature models. Each StructuralCheck contains a reference to a RelaxNG[2] schema that is tested against a node.

• The Interface part of a Test Step defines the feature model that represents the possible configurations of the Test Step instance that can be instantiated for the Test Step. As already described, the feature model both includes feature specific to the Test Step and defines references to Service instances.

[2]http://relaxng.org

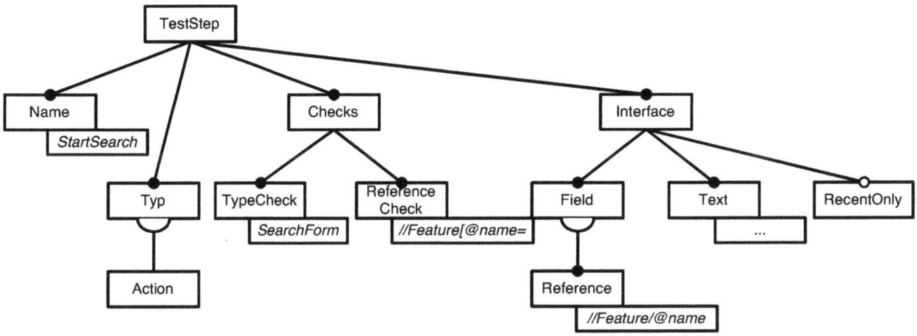

Figure 6.10: Feature model of the Interface elements of a Test Step

Figure 6.10 displays a feature model of a Test Step before checks are executed and the references are unfolded. The name of this Test Step is START_SEARCH. The Test Step represents the submission of a query to an IR system using a search form. This Test Step is of the type Action, meaning that it manipulates the state of a testee and causes it to change into a different context.

Two checks are defined for the Test Step: The TypeCheck specifies that instances of the Test Step can only be instantiated with instances of the SEARCH_FORM Service The ReferenceCheck checks for the existence of a feature with a specific name. The XPath used by the ReferenceCheck is truncated in the figure.

The Interface feature defines the interface that is available for this Test Step and that is represented as GUI elements by the editor. The Test Step in Figure 6.10 includes three elements:

- The feature group under the Field node defines a reference to the feature model of a Service instance. When this reference is unfolded, the subtree of the feature specified by the reference is copied into the feature model of the Test Step, replacing the reference.
- The Text feature is a part of the interface of the Test Step that allows the specification of textual information, represented by an attribute node.
- The RecentOnly feature represents a choice that can be made for this Test Step. The RecentOnly option can either be selected or de-selected for the Test Step.

6.3 Composition of the Application Test Model

The Application Test Model is a composition of the Application Feature Model, the Application State Model, and the Domain Test Model. The purpose of the model is to provide a representation of the SUT that is optimized as a basis for test modeling, that is the creation of Test Configurationinstances, using the MTCC editor.

6.3.1 Purpose of the Composition Process and the Application Test Model

The purpose of the composition process is the realization of the Application Test Model. It is the instantiation of a model whose dynamic parts describe the sequences of Test Steps that can be used to exercise and verify a testee while the structural parts of the Application Test Model describe the interface provided by each Test Step instance.

In contrast to the models from which it is created, the Application Test Model is a transient and automatically created model with the sole purpose to serve as an information source to the editor during test modeling. As a consequence, the Application Test Model is optimized to provide the editor with information in the least complex way feasible, but does not aim for good maintainability or the avoidance of redundancy.

6.3.2 The Composition Process

In the following we discuss the composition process in general as well as the derivation process used to realize the dynamic part of the Application Test Model.

As in the composition process used to derive the structural part of the Application Test Model, all Test Steps defined in the Domain Test Model are tested against all Service instances defined in the Application State Model. Figure 6.11 gives an overview of the composition process. If a Service instance and a Test Step are compatible, a state representing the Test Step instance is created and added to the FSM. Initially, the dynamic part of the Application Test Model is a graph that consists only of Context states: A start state and an end state. During the composition process, all Test Step instances that can be executed from a Context are added to the FSM. For each added Test Step instance, the following transitions are added to the FSM:

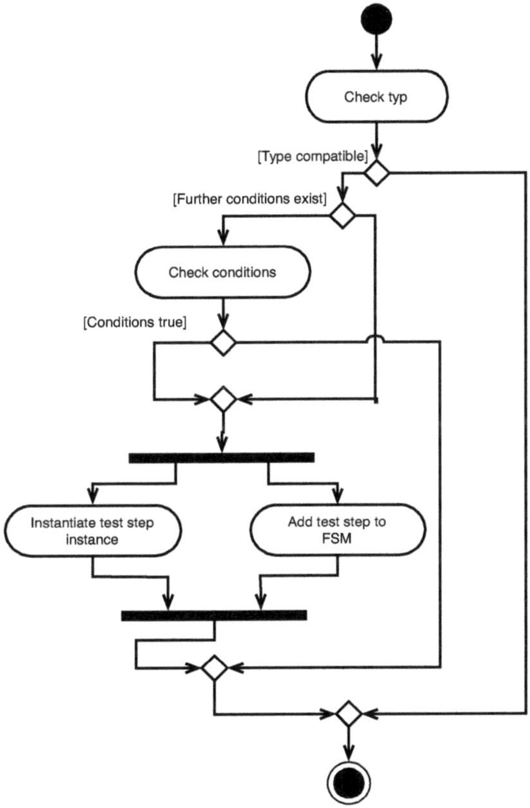

Figure 6.11: Flowchart of the MTCC Composition

1. A select Test Step transition from the context of the Service instance referenced by the Service to the Test Step instance is added to the FSM.

2. A execute Test Step transition is added to test the FSM for every Context state that represents a possible state of the testee after the execution of the Test Step instance is added to the FSM.

The resulting FSM is similar to a Application State Model where every Service state is replaced by one or more Test Step instance states. The number of states in the dynamic part of the Application Test Model is therefore higher than the number of states in the Application State Model.

The Context states that can be reached from the state of a Test Step instance are determined by the type of the Test Step instance: For Test Step instances that are either Assertions or SameStateActions, a transition to the initial Context from which the Test Step instance was selected is added to the FSM. For Test Step instances that have the type Action, transitions are added to all Contexts that are reachable from the Service instance of the Test Step in the Application State Model.

A number of Test Step instances, for example the comparison of variables, are not dependent on the Services of the system under test but represent functionality of the test system itself. In order to represent such Test Step instances in the same way as the Test Step instances that exercise or otherwise interact with the testee, MTCC uses a virtual Service instance, the DefaultService. The DefaultService is automatically included in every Context of a system. Every Test Step instance that must be available regardless of the Services provided in the current Context thus references an instance of the DefaultService. All instances of the DefaultService are identical, it does not define any interface.

6.4 Chapter Summary

MTCC uses feature models to represent the test-relevant functionality of a system and finite state machines to represent the behavior of the system.

The behavior of a system is described by a system-level model, the Application State Model. In order to represent the dynamic behavior of a system, the system is partitioned into Contexts. Each context is defined by a number of Services that are available for that Context. Contexts and Services are logically represented as states in a finite state machine. Transitions represent the invocation of Services and the resulting Context changes.

Feature models are used by MTCC both on the domain-level and on the system-level. The system-level feature models included in the Application Feature Model are specializations of the feature models defined in Domain Engineering that are included in the Domain Feature Model. Each feature model in the Application Feature Model represents an instance of a Service in one Context of the represented system.

In MTCC model composition, the Application Test Model, the Application Feature Model and the domain-level Domain Test Model are combined into a test-level model, the Application Test Model. The Application Test Model describes which Test Steps can be executed in the different Contexts of a system. In order to instantiate the Application Test Model, the Test Steps defined in the

Domain Test Model are tested for compatibility with the Service instances defined in the Application Feature Model. If a Service instance and a Test Step are compatible, a new Test Step instance is created. The static part of the Application Test Model describes the supported parameterization for Test Steps, the dynamic part describes the possible sequences in which Test Steps can be invoked to test a system.

7 Application of MTCC Models

This chapter examines the use of the Application Test Model for test construction. In order to facilitate the construction of tests by domain experts, the MTCC editor is used to allow the representation and configuration of the Application Test Model in a graphical user interface. MTCC supports the reuse of existing Test Configurations for multiple testees in a system family based on the transfer of Test Configuration instances in order to execute an abstract Test Configuration. To test, MTCC uses code generation to generate a concrete test script.

Section 7.1 examines the representation of the dynamic and structural parts of the Application Test Model in the MTCC editor. The editor represents tests as sequences of configurable Test Steps. Domain experts create tests by selecting and configuring the Test Step instances that represent the actions that are intended to exercise the testee. Test Step instance selection and configuration are done in the editor. The editor is a graphical user interface that represents specific features and subtrees of features by standard elements of graphical user interfaces like lists of values, test input fields or check boxes.

One purpose of the editor is the configuration of the Test Step instances defined in the Application Test Model and thus the removal of all variable elements of the feature model. While MTCC does not require or implement all specialization steps defined for feature models with cardinalities [CHE05c], it is important that specializations applied to Test Step instances are reversible to allow for the correction of mistakes. In order to support the reversal of specialization steps, MTCC introduces configuration nodes that support the explicit representation and manipulation of selection steps in MTCC feature models. Section 7.2 discusses the realization of the configuration activities for feature models in MTCC.

MTCC considers testees in the context of software system families and aims to support the reuse of Test Configuration models created based on the Application Test Model of one testee for other testees. Section 7.3 discusses the requirements for such reuse in MTCC and the techniques used to support the transfer of tests systems.

In order for tests to be transfered from one system to another, it is necessary that the target system supports the execution of all Test Steps used in the test in the same order in which they are used in the test.

Additionally, the Test Steps for the initial system and the target system must be compatible in the sense that the feature model of each Test Step instance in the target systems supports the configuration that was applied to the feature model of the corresponding Test Step instance for the initial system.

If more than one sequence of Test Step instances for the target system is compatible with the Test Step instances in the initial system, an automatic decision on the sequence that best represents the semantics of the original test is not possible. It is also not possible for MTCC to decide whether a test that can be technically transfered by MTCC really represents the intention of the test on the target system. For these reasons, test reuse in MTCC is a semiautomatic process.

The Test Configurations created in the test modeling process do not contain information about the technical details required for test execution. In order to exercise a system based on the tests, these models must either be interpreted by a runtime system or must serve as the basis for test code generation. Section 7.4 discusses the approach to test code generation used in MTCC. MTCC uses a template-based code generator that generates test cases for xUnit test runners. MTCC test cases depend on a library that contains implementations for all Test Step instances of a given system under test.

7.1 Representation of Test Steps in the Editor

The purpose of the MTCC editor is to support domain experts' formal modeling or programing skills in the specification of tests.

The test models created by the domain experts have to be formal in the sense that they support the generation of scripts without the need for further refinement or rework by domain engineers. In order to meet this requirement, the editor needs to fulfill the following conditions:

- The editor has to be able to represent the information contained in the Application Test Model in a form that allows a domain expert to express a testing intent in a MTCC test. The possible configurations of the MTCC editor have to be represented by GUI elements in an understandable way.
- The editor must be able to save tests in a format suitable of code generation.
- The editor has to be able to translate the interactions of a user with the GUI into specialization steps that configure the feature model of a Test Step instance. The specialization steps have to be expressed in a way that allows for changes when the configuration of a Test Step instance is changed in the editor.

In the following, we introduce the MTCC editor from a user's point of view. We examine the test modeling process as the selection and configuration of Test Step instances. Based on the external view on the editor, we examine the representation of feature models as GUI elements in Section 7.1.3.

7.1.1 Overview of the MTCC Editor

Figure 7.1 displays a screen shot of the editor. The user interface consists of three separate areas: A toolbar that gives access to the functionality provided by the editor is located at the top of the editor window. A list of the Test Step instances that are currently selected for a test can be seen in the left half of the editor, the right half of the editor consists of the GUI representation of the currently selected Test Step and a text area that explains the functionality of this test. The overall structure of the editor — a list that allows the selection of elements in the left part of the GUI and a detailed representation of the selected element in the right part — has been chosen for MTCC because it is frequently used in programs in other domains, for example, in email clients. It is therefore likely that users are familiar with this representation.

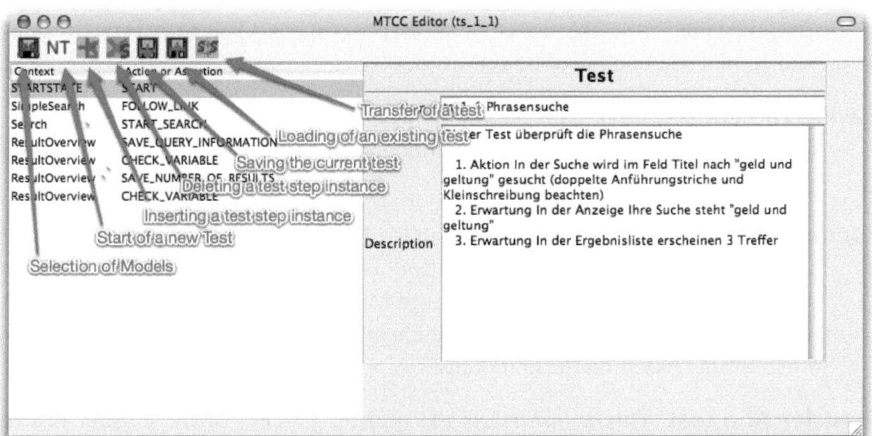

Figure 7.1: Screenshot of the start page of the MTCC editor

The toolbar gives access to the functions of MTCC such as loading and saving of tests and the insertion and the deletion of Test Step instances. All functions of MTCC that can be selected with the toolbar are also available in a menu not displayed in Figure 7.1. Starting from the left, the following functions or actions can be invoked from the toolbar:

Selection of models This button opens a dialog that allows the selection of the currently used Application State Model, Application Feature Model and Domain Test Model. All three models must be available as XML files in the file system. Once the models are loaded by the editor, a new Application Test Model is created in the composition process discussed in Section 6.3.

Start of a new test This action starts a new test for the currently selected Application Test Model. The Test Step instances that are currently loaded into the editor are lost.

Inserting a Test Step instance This action inserts a new Test Step instance into the currently loaded test. A dialog allows the selection of a Test Step instance from a list, the Test Step instance is then inserted after the Test Step instance that is currently selected. Which Test Step instances can be inserted, is determined by the Application Test Model.

Deleting a Test Step instance This action removes the currently selected Test Step instance from the list of Test Step instances. The first Test Step instance in a test can never be removed.

Saving the current test This action opens a dialog that allows the selection of a file in which the current test can be saved as a Test Configuration in a XML format.

Loading of an existing test This action loads an existing test in the editor and changes the currently used Application Test Model by loading the Application State Model, Application Feature Model, and Domain Test Model that were used for the construction of the test and composes a new Application Test Model based on these models.

Transfer of a test This action starts the transfer process of the current test to another system, the target system of the test transfer is selected from a list of all modeled systems for the current system family.

The list of Test Step instances includes all Test Step instances that are part of the current test in the order in which they will be applied to the testee. Each Test Step instance is represented by the type of the Test Step and the name of the Context of Test Step. If more than one instance of a Service exist for the

Context, the name of the Service instance that the Test Steps invoke is displayed in parentheses.

In Figure 7.1, the currently selected Test Step is the START Test Step. This Test Step instance is part of every MTCC test. It represents the start state of the FSM and cannot be removed from the test. This Test Step instance does not represent an interaction with the testee but allows the definition of metadata for the test, specifically a description and a title.

7.1.2 Construction of Tests from Test Step Instances

A test in MTCC is a sequence of configured Test Step instances. One part of test modeling is therefore the selection and combination of those Test Step instances that represent the interaction between a user and the testee that is to be tested.

Figure 7.2 illustrates the dialog used for the selection of Test Steps. The Test Steps presented for selection are those that are available for the current Context based on the information included in the Application Test Model.

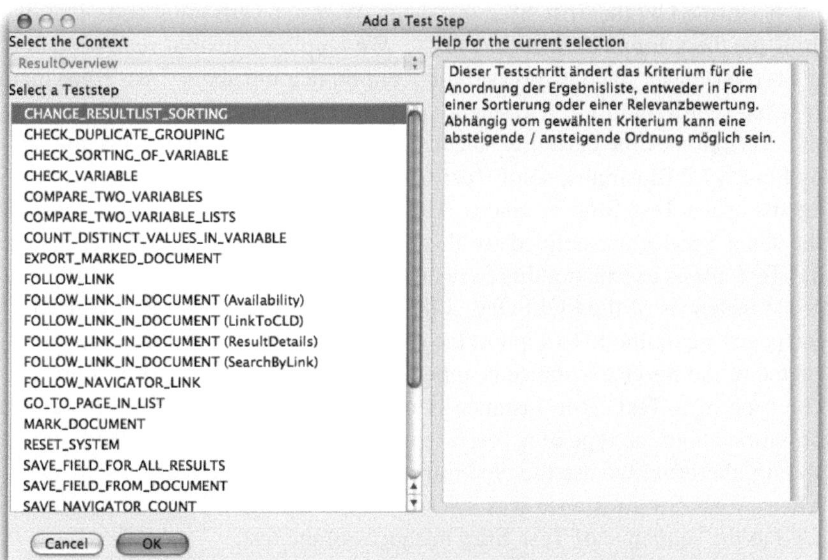

Figure 7.2: Dialog for the selection of Test Step Instances

The graphical user interface of the dialog used for Test Step instance selection is partioned in two parts. The left part of the dialog is a list of all Test Step instances available in the current Context. If the system can be in one of multiple Contexts as the result of the execution of the previous Test Step, the current Test Step can be selected from a Drop Down list. The right part of the dialog displays a help text for the currently selected Test Step. The selection of a Test Step instance is done in two steps: First a Context is selected, then a Test Step instance from the Context.

In Figure 7.2, only Test Step instances from one Context can be selected, the control for the selection of Contexts is therefore grayed out.

In general all Contexts that are potential successor contexts of the Service instance of the previous Test Step instances can be selected. Since the Application Test Model does not contain information about the semantic differences of the different Contexts, the selection of a Context is based only on the expectations of the user.

In the dialog displayed in Figure 7.2, a Test Step instance is selected from a list of all available Test Step instances sorted in alphabetic order by the name of the Test Step. The name of each Test Step serves as a minimal description of the action performed by the Test Step instances. We use a verb-noun combination for each of the Test Steps defined in MTCC. We employ a textual representation of Test Step instances since it allows easier communication about Test Steps than an icon based representation. The selection of the right Test Step is supported by a short description of the currently selected Test Step in the right half of the dialog.

As Figure 7.2 illustrates, some Test Step instance are included more than once in the list of the Test Step instances. The reason for this is that multiple instances of the same Service are defined for this Context, this leads to multiple instances of the Test Steps exercising this Service For the system represented in the figure, multiple instances of the FOLLOW_LINK Service representing HTML Links between pages are defined. To support the selection of the correct Test Step instance, the name of the Service instance is appended to the name of the Test Step.

The type of a Test Step instance is not displayed in the list. We argue that information about the type of a Test Step instance does not support the user in test modeling, therefore we use the type information only internally.

After a Test Step instance was selected in the dialog, it is inserted into or appended to the sequence of Test Step instances of the test.

When a Test Step instance is added or removed from the test, the sequence of Test Steps can become invalid. For instance, when a Test Step instance that submits a query to a Digital Library is removed from the list of Test Steps, the following Test Step instances that rely on the fact that the search Test Step caused the system to change, cannot be executed. In order to detect situations where the sequence of Test Steps is not valid, MTCC checks the list for correctness after each addition or deletion. If the sequence is invalid, that is, if the sequence of Test Step instances cannot be reproduced with the dynamic part of the Application Test Model, the name of the first invalid Test Step instance in the list is displayed in red. The validity can be restored by adding or removing the correct Test Step instances.

7.1.3 Representation of a Test Step Instance by GUI Elements

The second requirement of the editor beyond the construction of valid Test Step instances is the representation of the feature models of these Test Step instances in a form that allows their configuration by domain experts. In MTCC, the mapping of feature models to GUI elements and thereby the representation of feature models in a suitable form for domain experts is done by the GUIBuilder class and a number of subclasses of the Buildlet class.

These classes select the GUI elements that represent a given feature or feature subtree, build the GUI representation and integrate it with the GUI for the other features of the interface of a Test Step instance.

In contrast to other approaches for the creation of GUIs for feature models, the approach used by MTCC is not solely based on the structure of the feature model but takes into account the concepts that are represented by the feature models and the conventions for the representation of these concepts in the respective application domain.

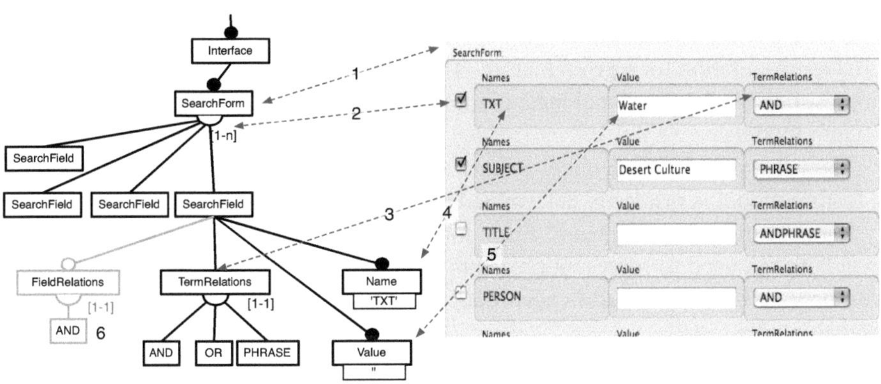

Figure 7.3: Mapping of features to GUI elements

Figure 7.3 illustrates the mapping of the nodes of a feature model to GUI elements in the MTCC editor. The following aspects of the mapping are of special relevance:

1. The hierarchy of the feature model is mirrored by the hierarchy of the GUI elements. The SearchForm and SearchField nodes of the feature model, for example, correspond to the hierarchy of the GUI containers used to display them.
2. The optional children of the SearchForm feature group are represented by Check Boxes.
3. If a feature group has a maximal cardinality of 1 and thus allows only the selection of a single child, and if none of the child nodes has children, this feature group is represented by a Drop Down list.
4. An attribute that is already configured in the Application Test Model and thus has no remaining variability, is represented as a static test panel.
5. Attributes that are not configured in the Application Test Model are represented by a text input field.
6. Feature groups with only one child and a cardinality of [1,1] — and thus no variability — are not represented in the GUI.

Figure 7.4 illustrates the relationship of the classes GuiBilder and Buildlet and their subclasses. The GuiBuilder class coordinates the mapping of feature models to GUI elements and identifies the subclass of the Buildlet class that can build the GUI representation for a given feature. The instances of the subclasses of Buildlets are responsible for the actual instantiation of the GUI elements for a subtree of the feature model.

The behavior of a GuiBuilder instance and therefore the GUI representation of a feature model is primarily determined by its associated Buildlet objects.

MTCC differentiates generic from domain-specific Buildlets. Domain-specific Buildlets are used to represent concepts from the application domains under test that have to be represented in accordance with the conventions of the application domain under test.

An example from the application domain of Digital Libraries are complex search forms. Domain experts often expect that such concepts are represented by a specific arrangement of GUI widgets. In order to support such convention in MTCC, domain-specific Buildlets are used.

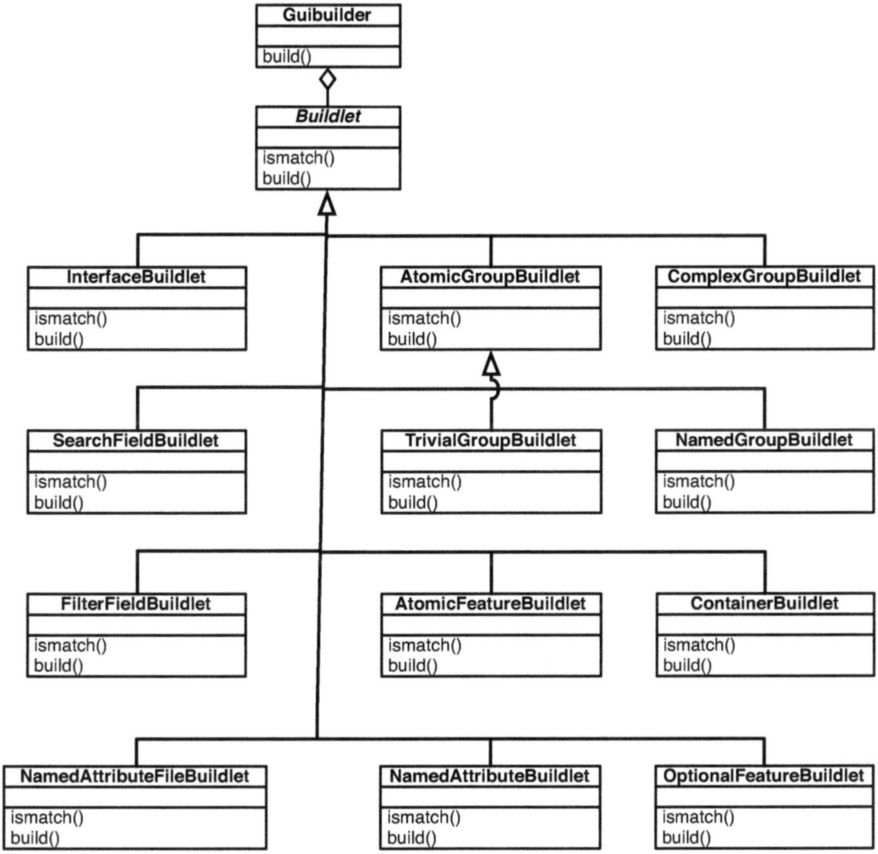

Figure 7.4: Classes used in the GUI representation of feature models

Generic Buildlet objects are not specific to the application domain and instead build a representation of feature models based on the structure of a feature model. Factors that determine the representation include the node type, the cardinality and in more complex cases the structure of a subtree of features.

The build method of the GuiBuilder class is called when a Test Step instance is selected in the editor. The following activities are performed in order to map a feature model to a tree of GUI elements:

- The list of all Buildlet objects registered with the GuiBuilder instance is traversed until an object is found that supports the mapping of the current feature node and potentially its sub nodes into GUI elements. In order to check the suitability of a Buildlet, its ismatch() method is called. Among the aspects considered by the ismatch() methods of different Buildlets are the name of the feature, its type, its cardinality and the type and cardinality of its children.

  The List of Buildlet objects is traversed sequentially. The first Buildlet, for which the execution of ismatch() returns true, is used for the representation of the feature. Generally, the most specific Buildlets are tested first, while the more generic ones are considered last. We discuss the Buildlet objects used in the Context of this work and the sequence in which these objects were arranged in a later part of this section.

- After a suitable Buildlet instance has been found, its build() method is called with the current node of the feature model as its parameter. This method instantiates the GUI elements that represent the current node and potentially its children. In order to connect interactions with the user interface with the configuration of the feature model, callbacks that apply specialization steps to the feature model are registered.

  If a feature model has children that are not mapped to GUI elements by the currently active build() method, the Buildlet either calls the build() method of another Buildlet directly or delegates to the GUIBuilder that determines a suitable Buildlet by the process described above.

  The result of a call to the build() method is an object from the hierarchy of GUI classes used to implement the editor. This GUI object will often be a container class that contains the GUI elements of children that were mapped by other Buildlet instances.

The implementation of the MTCC approach prepared for this work uses four domain-specific and eight generic Buildlet classes. We first discuss the domain-specific Buildlet classes in the order in which they are tested against a feature model:

InterfaceBuildlet This Buildlet constructs the container object that wraps all GUI elements used for the representation of a Test Step instance interface.

SearchFieldBuildlet This Buildlet builds a representation of a single field of a search form. As Figure 7.5 illustrates, the representation of a search field is similar to the GUI elements used by the systems within the system family itself. Such a representation would not be possible on the basis of generic Buildlet objects alone.

FilterFieldBuildlet Analogous to the SearchFieldBuildlet, this object represents a filter used in a search form. Figure 7.5 displays the representation of a filter in the editor in comparison to the GUI used by the Sowiport Digital Library.

NamedAttributeFileBuildlet This Buildlet represents a single attribute node that is used to store file names as its value. While this Buildlet is not strictly specific to the application domain of Digital Libraries, it is domain-specific insofar in that it represents the attribute based on its intended usage, not its structure.

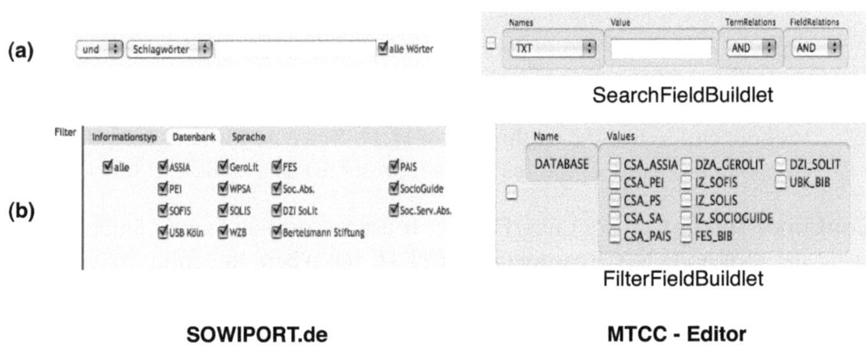

Figure 7.5: Representation of domain-specific features in the GUI and in the MTCC editor

The following generic Buildlet objects are used in MTCC. Figure 7.6 displays a part of the graphical user interface of the editor and relates the utilized GUI elements and their relation to the generic Buildlet responsible.

TrivialGroupBuildlet This Buildlet is used to represent feature groups with children that are leaves in the feature diagram. In cases where the cardinality of the feature group is [1,1], the feature group is represented by a GUI element that allows the selection of a single value from a list. Otherwise each child feature is represented by a Check Box.

AtomicGroupBuildlet This Buildlet is a subclass of the TrivialGroupBuildet for feature groups with a single child node. When the cardinality of the feature group is [0,1], the feature group is represented by a Check Box with one option that represents the single child of the feature group. If the cardinality of the feature group is [1,1] and thus the child node is mandatory, it is selected automatically and the feature group is represented by a static text label.

ComplexGroupBuildlet This Buildlet is used for the representation of feature groups with non-trivial child nodes. Depending on the cardinality of the feature group, children are either represented by a Check Box for each child, or in cases where the cardinality is [1,1], by a radio box. The mapping of child nodes of the feature group to GUI elements is delegated to other Buildlets.

NamedAttributeBuildlet This Buildlet is used for the representation of feature nodes with an attribute node as its only child node. The feature is represented by a text input field. The name of the feature that the attributes belongs to is displayed in the border around the input field.

AtomicFeatureBuildlet This Buildlet is used for the display of a single feature without children and a cardinality of [1,1]. Since no further specialization is possible, the name of the feature is represented as a static text label.

ContainerBuildlet A ContainerBuildlet is used to display feature models where the root node has a cardinality of [1,1], but where the child nodes can be configured. Such features are represented by a container object of the GUI library used for the editor. The container object is not displayed in the GUI, it only serves as a holder for GUI elements.

OptionalFeatureBuildlet This Buildlet is used for the display of optional features with a cardinality of [0,1]. The representation of the children of this feature is delegated to another Buildlet.

7.2 Configuration of Test Models

The MTCC editor must not only support the display of Test Step instance feature models but also allow domain experts to interact with these models. Specifically, it has to translate user actions into specialization steps that configure the feature model.

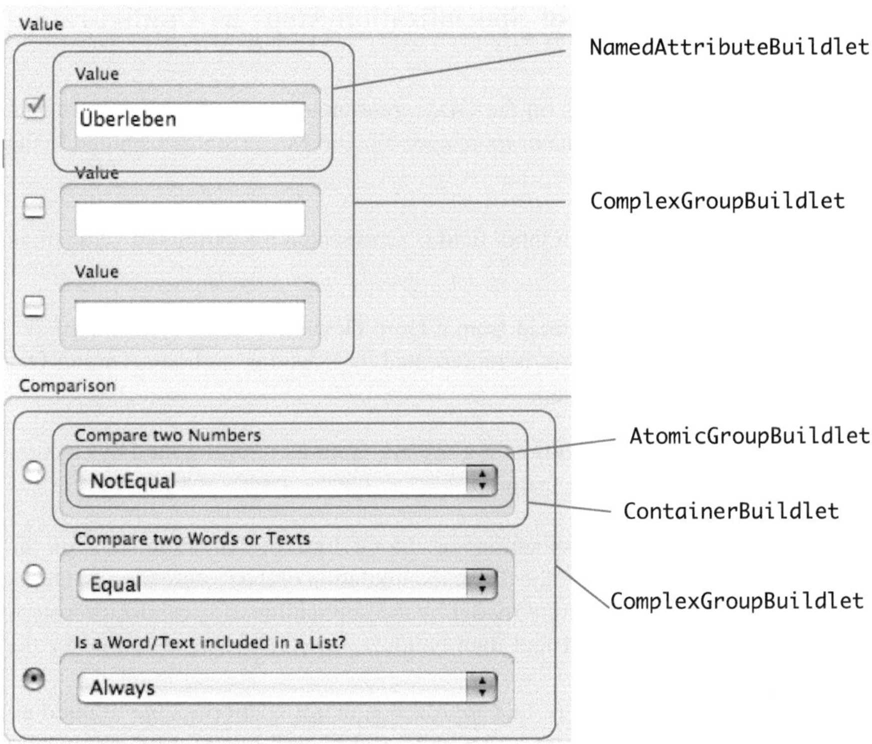

Figure 7.6: Part of the MTCC editor with the Buildlets used

The MTCC editor uses event handlers for every control that represents a part of the feature model of a Test Step instance. Each event handler holds a reference to the feature node represented by the GUI and calls methods on its interface to reflect the configuration of the Test Step instance as specified with and represented by the GUI.

On the level of the feature model, the configuration of a Test Step instance is represented by configuration nodes that are appended to the nodes of the feature model which are the subject of configuration. Each configuration node represents a single specialization step. The representation of specialization steps has a number of advantages compared to the direct application of specialization steps and thus the manipulation of the feature model. For instance, it supports changes to already specialized nodes.

7.2.1 Representation of Specialization Steps as Configuration Nodes

Every action that a user takes on the GUI representation of a Test Step instance results in the application of one or more specialization steps that are applied to the feature model:

- The input of text into an input field is represented by setting the value of an attribute node.
- The selection of an element from a Drop-Down list or the selection of elements from a Check Box is represented by selecting and de-selecting features from a feature group.
- The selection or de-selection of a solitary features is represented by changes to the cardinality of the feature.

The obvious way to express actions on the GUI as specialization steps on the feature model of a Test Step is the direct manipulation of the feature model. Direct manipulation changes the feature model by the application of specialization steps with every action of the GUI and then builds a new GUI representation for the changed feature model.

A significant disadvantage of this approach is that it is not possible to undo already applied specialization steps. A specialized feature model does not contain information about its structure before the specialization occurred. For example, if a solitary feature with a cardinality of [0,1] is selected and thus the cardinality is changed to [1,1], it is not possible to discern its cardinality previous to configuration. For the representation of feature models in the MTCC editor, this loss of information is not acceptable. It is not possible to change the configuration of a Test Step instance once a specialization step was applied to the feature model, since the initial state of the feature model is lost.

Another disadvantage of the direct manipulation approach concerns the consistency of the GUI. Since each specialization step changes the feature model underlying the GUI, the GUI elements also change with every interaction with the GUI. For example, once a value has been selected from a Drop-Down list, the list itself would be removed from the GUI.

In order to support changes to the specialization steps that were applied to a feature model and to avoid changes to the overall structure of the GUI representations of Test Step instances, at least two different approaches are available: The utilization of multiple versions of a feature model or the representation of specialization steps as distinct entities, similar to the Command Pattern [GHJV95]. We choose the latter approach for MTCC: Specialization steps are represented as nodes in the feature model.

7.2.2 Implementation of Configuration Nodes

Figure 7.7 displays the class diagram of the feature object used in MTCC together with the classes used to represent the specializations that can be applied to each node.

To support the representation of specialization steps as nodes in the feature model, two class hierarchies are used. One class hierarchy consists of *Mixin* [BC90] classes used to extend the functionality of the original feature classes. A Mixin is an abstract class that defines methods that are added to the functionality of classes that inherit from the Mixin. Mixins only serve to provide functionality to subclasses, an is-a relationship between a subclass and the Mixin does not exist.

The second class hierarchy includes classes representing the configuration nodes. The Mixin classes extend the functionality of the feature nodes with methods for the creation, removal, and querying of feature nodes as well as functionality that allows the configuration nodes to be applied to the feature model.

These methods are used by the event handlers registered for actions on the user interface. The decision to use Mixin classes as the basis of the implementation was made to decouple the configuration mechanism from the basic classes used for feature modeling. The subclasses of BaseConfigurationMixin do not have any class attributes. They only extend the available methods of features, feature groups and attributes.

The configuration nodes store the specialization steps that exist for a given node. Every configuration node is associated with one node of the feature model. Figure 7.8 gives an example for the relationship of feature nodes and configuration nodes. Not all specialization steps defined for feature models with cardinalities are implemented in MTCC, the cloning [CHE05c, CHE05b] of features for instance is not supported.

When the feature model of a Test Step instance is loaded into the editor, a default specialization for each feature is instantiated and appended to the feature. This default configuration corresponds to the configuration of the GUI elements in the user interface.

MTCC implements the following configuration nodes:

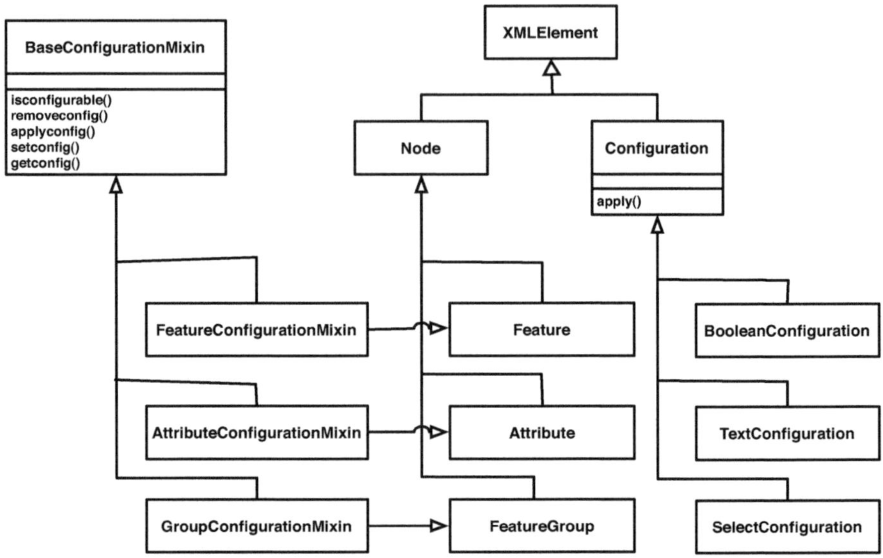

Figure 7.7: Class hierarchies of features and configuration nodes

BooleanConfiguration This configuration object is used for the selection and de-selection of features with a cardinality of [0,1]. Each instance of BooleanConfiguration stores a single Boolean value that represents the configuration of the node. A cardinality of [0,0] is represented by the Boolean value false. A configuration of [1,1] is represented by the value true. The default configuration of the value is false.

TextConfiguration This configuration node is used to store values that are assigned to attribute nodes. MTCC does not use a type system, the value currently stored by the attribute is therefore always represented as a string. The default value is an empty string.

SelectConfiguration This configuration object is used to represent the selection of nodes from the available children of a feature group. All legal cardinalities for feature groups are supported. The configuration object stores the selection of values as a list of Boolean values. The default configuration for this node selects the minimal number of children that must be selected according to the cardinality of the feature group.

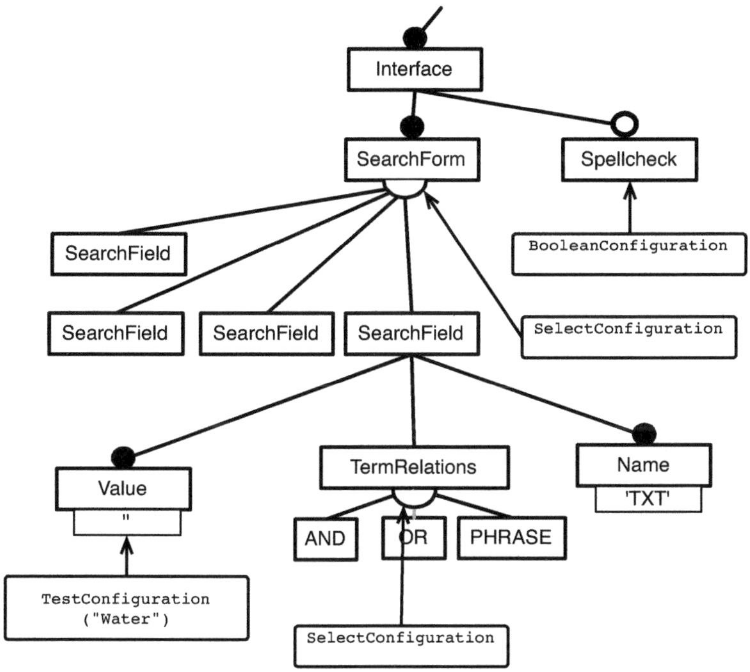

Figure 7.8: Feature model with configuration nodes

7.2.3 Application of Configuration Objects

When configuration nodes are used, the transformation of a feature model according to the specified specialization steps only takes place when the configuration nodes are applied to the feature model.

In addition to the transformation semantics for feature models with cardinalities as described in Section 3.2, MTCC also defines a set of transformation rules that conserve the position of nodes within the tree of the feature model. Figure 7.9 displays the feature model from Figure 7.8 after the configuration nodes have been applied. The absolute position of many nodes in the feature model has changed, for example, all fully specialized feature group nodes are no longer part of the model.

Figure 7.10 displays the feature model after the configuration was applied using the MTCC semantics. The absolute positions of all remaining nodes is unchanged.

While MTCC preserves nodes that do not add information to the feature model and are thus redundant, the transformation displayed in Figure 7.10 facilitates the comparison of feature models before and after the application of a specialization step. MTCC uses this property when testing the reusability of test code.

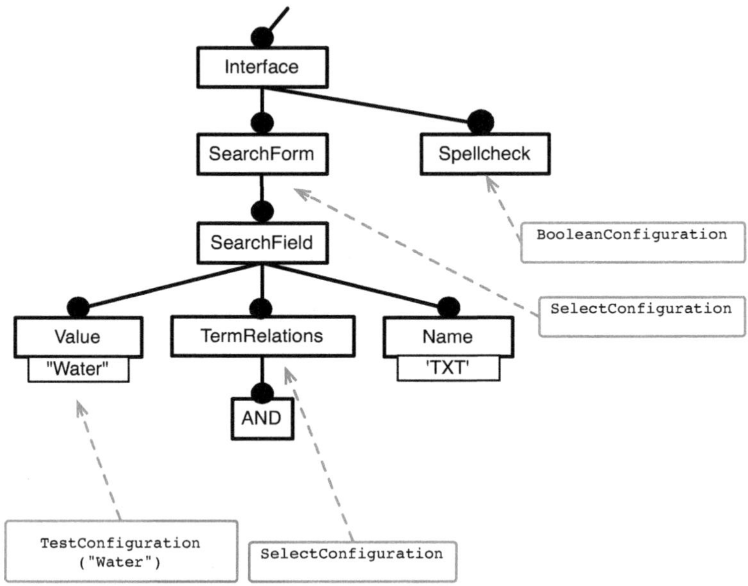

Figure 7.9: Feature model after the application of its configuration nodes

7.3 Reuse of Tests

MTCC approaches the reuse of tests for different systems by providing functionality for the transfer of Test Configuration instances creates for one Application Test Model to a target Application Test Model. A target Application Test Model is compatible with a Test Configuration if it supports the Test Step sequence specified in the dynamic part of the Application Test Model and if the Test Step instance of the target Application Test Model can be configured like the Test Step instances of the test.

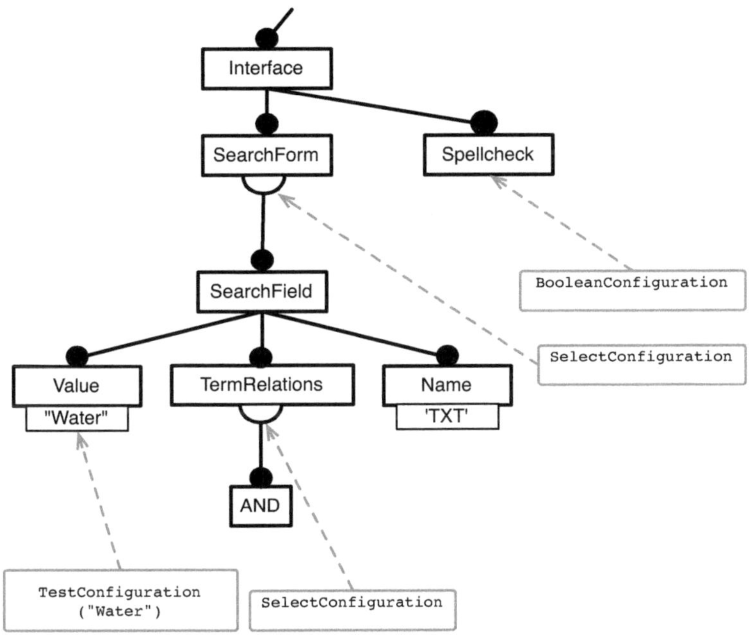

Figure 7.10: Feature model after the structure-preserving application of its configuration

7.3.1 Transfer of Test Step Sequences

As described in Section 6.1, MTCC represents tests as sequences of configured Test Step instances. Every Test Step instance represents an action or a check for the specific interface of the system under test.

Which sequences of Test Step instances are available for a system, is determined by the dynamic part of the Application Test Model. A sequence of Test Steps can only be applied to a testee if the FSM can produce a sequence of Contexts and of Test Step instances of the appropriate type.

This condition can be formalized as follows: If only the types of Test Step instances are considered then a test can be transfered to another system when the FSM of the dynamic part of the Application Test Model for the target system can be transformed into an acceptor FSM whose states are the Test Step instances for the system and in which a transition between the Test Steps exists if a Test Step instance in the Application Test Model can be reached via any Context node. Figure 7.11 gives an example for a finite state machine in the acceptor representation. The FSM accepts a test that consists of the Test Steps $Start, A, C, D, E$. The test $Start, A, B, C$ is not accepted.

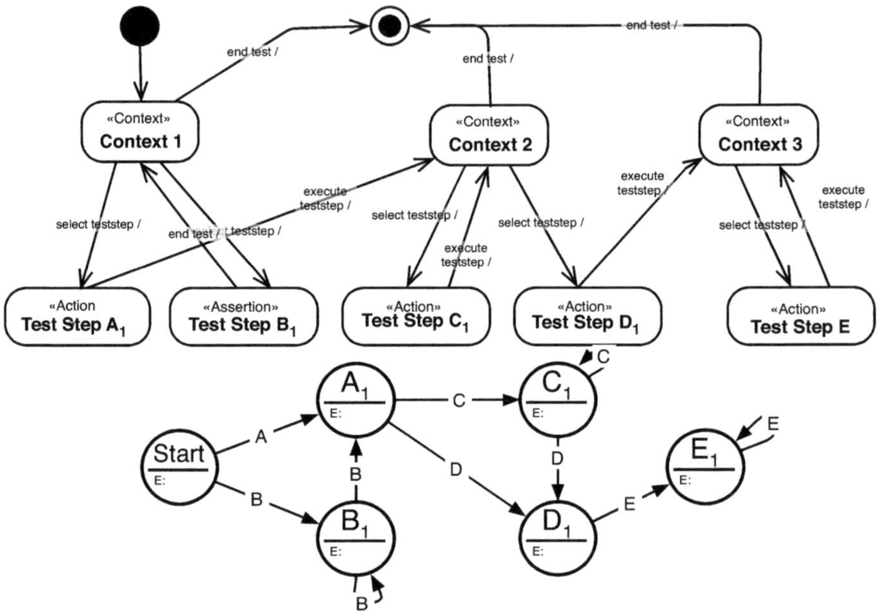

Figure 7.11: Acceptor for Test Step sequences based on the Application Test Model

As we discussed in Section 6.1, since MTCC supports more than one instance of a Service in one Context, it is possible that more than one instance of a Test Step exists for one Context. MTCC also supports Test Step instances that are nondeterministic insofar that they can transfer the SUT into one of multiple Contexts. Both these properties lead to a situation where one sequence of Test Steps can be expressed by different Test Step instances.

As Figure 7.12 illustrates, an acceptor in MTCC is not deterministic for the reasons described above. The test $Start, A, C, D, A$ can be accepted by both the sequence $Start, A_1, C_1, D_1, A_1$ and $Start, A_1, C_2, D_1, A_2$. An unambiguous mapping of the Test Step instances of a given test to the Test Step instances of a target system is not available for all testees.

Since it is not possible for an automated system to decide which of the multiple sequences of Test Step instances best represents the intention of that test, MTCC

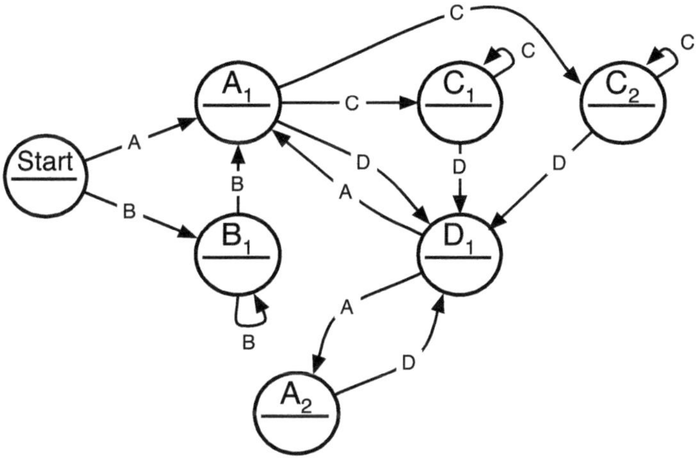

Figure 7.12: Non-deterministic acceptor in MTCC

uses an semi-automated approach to test reuse. MTCC builds a list of all candidate sequences that can represent the test on the target system. When this list is build, the user selects the sequence that is the best representation of the intended behavior or, if none of the candidate sequences is suited, aborts the test transfer process.

MTCC represents the different candidate sequences as a tree of Test Step instances whose root is the start state of the dynamic part of the Application Test Model. Each traversal through the tree from the root to a leaf represents one trace through the Test Step instances of the test.

MTCC uses a dialog that displays the tree of candidate sequences to the user. Figure 7.13 displays an example of this dialog: For every Test Step instance in the test, a Drop-Down list of candidates is displayed. It displays for each Test Step instance the name of the Service instance exercised by the Test Step instance and its respective Context.

The dialog reflects the tree structure of possible traces through the system. The Test Step instance available for selection at each moment depends on the Test Step instance selected for the previous step in the sequence.

In order to transfer a test as described above, it is not sufficient that a sequence of Test Step instances of the appropriate type can be found. It is also necessary that the configuration of each Test Step instance can be transfered to the corresponding Test Step instance of the target.

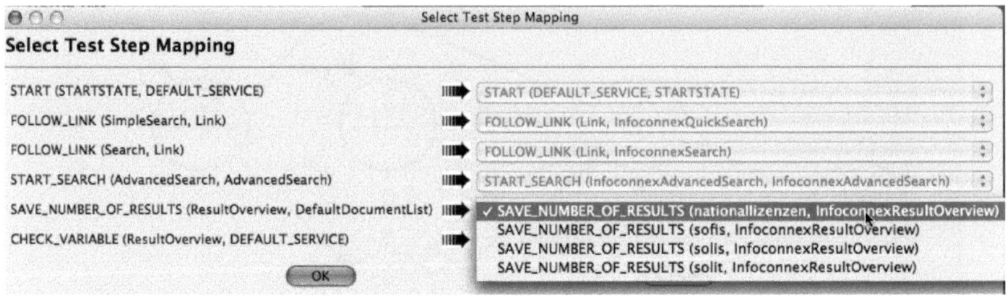

Figure 7.13: Dialog for the selection of Test Step instance sequences

7.3.2 Transfer of Test Step Configurations

In order to transfer a test from one system to another, it is necessary for the target system to have all features that are needed for the execution of the test. In MTCC, tests are transferable if all Test Step instance are transferable. A Test Step instance is transferable if the Service instance of the target system supports all features that are needed for the execution of the action or check that is represented by the Test Step instance. On the implementation level, MTCC considers a Test Step instance to be transferable when the configuration of the Test Step instance can be reproduced for the Test Step instance of the target system. In more detail, MTCC considers the feature models of two Test Step instances compatible if the feature model of a configured Test Step instance after the application of the configuration nodes is a rooted subtree of the the feature model of the target Test Step instance. Figure 7.14 displays a feature model and a target feature model. The models are considered compatible since the left model is a subtree of the right model.

In order to test the compatibility of two Test Step instances, the following activities are carried out:

- The Test Step instance to be transfered is copied. The configuration objects of the copy are applied as described in Section 7.2.3. The application of the configuration nodes removes all feature nodes that are neither selected by a user nor included in the feature model because the editor set a default value for the node.
- The feature models of both Test Step instances are canonicalized by recursively sorting all children for every node in the feature model. The sorting process considers the type of a node, its name and cardinality and in the case

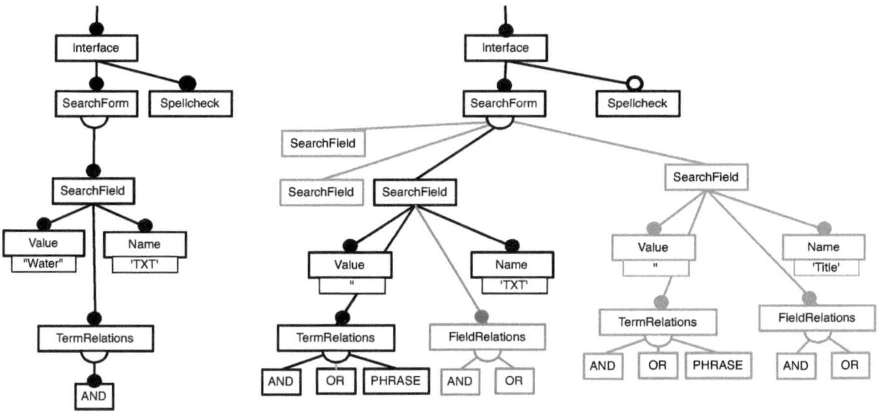

Figure 7.14: Two compatible feature models

of attributes, its value. The purpose of the canonicalization is to facilitate the later comparison of feature models.

- The configured Test Step instance is tested against the target feature model to determine if it is a subtree. The Test Step instance is a subtree for every node present in the model there is an equal node in the model for the target system. Since both feature models are canonicalized, the upper bound of the number of comparisons necessary to check for a subtree relationship is equal to the number of nodes in the target model. Two nodes are considered equal if their type, name, cardinality and value are either equal or if a specialization action exists that can be applied to the target node that makes them equal.

When the user has selected a sequence of Test Steps from the set of candidate sequences that are compatible to the Test Step instances of the original test, the target Test Step instances are configured. The goal of this configuration is to apply specializations to the feature model of the target Test Step instance in such a way that the selected nodes of the target Test Step instance are equal to the selected nodes of the original Test Step instance feature model. We call this activity configuration transfer. In order to transfer a configured Test Step instance, the configuration nodes of the original Test Step are applied to the feature model. The resulting subtree of the feature model is then used as a template for the target feature model. The nodes of the target model that correspond to the nodes of the subtree are configured such that all subtree nodes are selected in the target model. If the number of children of the original and the target node differ, the cardinalities of feature groups are adapted as necessary.

One consequence of the method described above is that the target Test Step may include features that are not present in the original model. The FieldRelation node in Figure 7.14 is an example for a feature that is only present in the target model. Since MTCC cannot determine if such features change the semantics of a test, the results of the test transfer process must be verified by a domain expert familiar with the system.

7.3.3 Limitations of Test Reuse

Some limitations apply to the approach to test reuse discussed in this section. When MTCC builds the set of candidate sequences of Test Step instances, it only considers sequences as compatible that contain exactly the same Test Steps in exactly the same order as the original system. In practice, these requirements are too severe. The order, in which assertions are executed on a system, for instance, are not always relevant to the behavior of a test.

One potential weakness concerning the transfer of configurations is that features that are explicitly de-selected are not considered when the configuration is transfered. In practice, this has little consequence: Optional features are by default not selected in the MTCC editor. This means that a user will not often de-select features that he or she does not wish to be used in the test.

7.4 Test Execution Based on Test Configuration Instances

The Test Configuration instances modeled by a domain expert are by themselves not sufficient for testing as they lack detailed information about the implementation of the testee and the test runner used for test execution. In order to test a system, MTCC uses a code generation approach that generates test cases from abstract models and templates.

The test cases generated by MTCC utilize the Test Adapter as a platform that provides functionality needed to exercise a SUT with a specific test runner. Figure 7.15 displays the artifacts relevant for test code generation in MTCC. In the following we examine these artifacts, their purpose and their interaction.

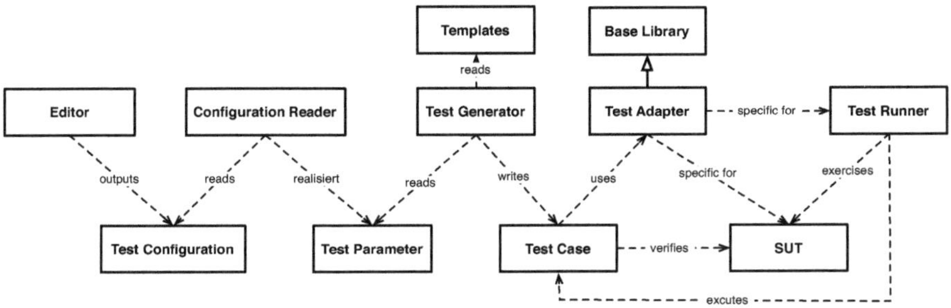

Figure 7.15: Artifacts used in test case generation

7.4.1 Transformation of Test Configuration Instances

Before test cases are generated from Test Configuration instances, each instance is transformed into an intermediate format. The purpose of the Configuration Reader is to transform a Test Configuration instance created with the MTCC editor into a simple representation that is used in the further generation of test code. The feature model of each configured Test Step instance is transformed into a Test Parameter object. Every Test Parameter object is implemented as an associative array with entries that represent the same information that is contained in the feature model of a configured Test Step.

Most entries in the Test Parameter object are strings, other possible data types for values include lists, numbers or other associative arrays. Listing 7.1 gives an example for a Test Parameter object in Python notation.

```
para = {
        'searchfields': [
                {
                        'term_relation': 'AND',
                        'field_relation': 'AND',
                        'name': 'TITLE',
                        'value': u'Frauen␣Maenner'
                },{
                        'term_relation': 'AND',
                        'field_relation': 'AND',
                        'name': 'YEAR',
                        'value': '1990-2005'
                }
        ]
}
```

Listing 7.1: Representation of a Test Parameter object as an associative array

One purpose of the Configuration Reader is the validation of Test Configuration instances, especially regarding the question if all necessary information needed for the Test Step instance is included in the model. The primary reason for the transformation into Test Parameter objects is the simplification of the data format used to express the semantics of the test. An associative array is used for the simplified data format since its availability can be assumed for any programming language.

This is relevant if the MTCC code generator and the test runner used for test execution are not written in a common language.

The actual transformation of the Test Configuration instances into Test Parameter objects is done by classes implemented at the system family level and is thus independent from the concrete system under test or the test runner used to execute tests.

7.4.2 Test Case Generation

MTCC assumes that test cases are generated for an xUnit implementation. Specifically, we assume that a test is represented as a test case that includes a series of method or function calls. It is assumed that a test fixture can be initialized by setUp methods/functions and that individual tests are organized into test suites which are automatically executed.

Assuming an object-oriented target language of test generation, every MTCC test is represented by a source code file with a single test class that contains a single test method. While this approach results in many individual files, it facilitates the selection of tests for execution.

The structure of a test method in MTCC is a sequence of method calls to the Test Adapter library that implements the functionality needed to exercise the system under test with the current test runner. Each method call represents the execution of a Test Step instance and is called with the Test Parameter object that represents the configuration for that Test Step and an object that stores information about the current state of test execution, for example, internal variables.

The Test Adapter contains an implementation of every Test Step instance for a specific combination of SUT and test runner. Listing 7.2 is an excerpt from a generated test case in Python. Some details of the code are omitted for clarity, the Test Parameter instances, for example, are not included in their entirety.

From the perspective of model-driven software development as described by Stahl and Voelter [SV06] and illustrated in Figure 7.16, the Test Adapter is a

```python
import sowiport_executor_selenium as runner

class PortalTest(unittest.TestCase):
    def test(self):
        context=Context()

        para = {...}
        self.failUnless(runner.start(context,para))

        para = {...}
        self.failUnless(runner.follow_link(context,para))

        para = {...}
        self.failUnless(runner.follow_link(context,para))

        para = {...}
        self.failUnless(runner.set_filter(context,para))

        para = {...}
        self.failUnless(runner.start_search_advanced(context,para))

        para = {...}
        self.failUnless(runner.save_navigator_count(context,para))

        para = {...}
        self.failUnless(runner.end(context,para))

if __name__ == "__main__":
    unittest.main()
```

Listing 7.2: Generated test for the programming language Python

platform for the generated code. For test scripts created by MTCC, this platform contains implementation details about the execution of test cases. In this context, the generated test cases are schematic code and the Test Configuration instances correspond to the application model.

MTCC does not require or assume the use of any specific approach to code generation. We argue nevertheless that the simple structure of the test code generated with MTCC are best generated with a simple approach like the *Templates and Filtering* [SV06] method.

More complex approaches [Voe03, Völ05, Völ06b], while advantageous for complex code generation scenarios, are over-specialized for the generation of tests that are basically linear sequences of method-calls without any control flow.

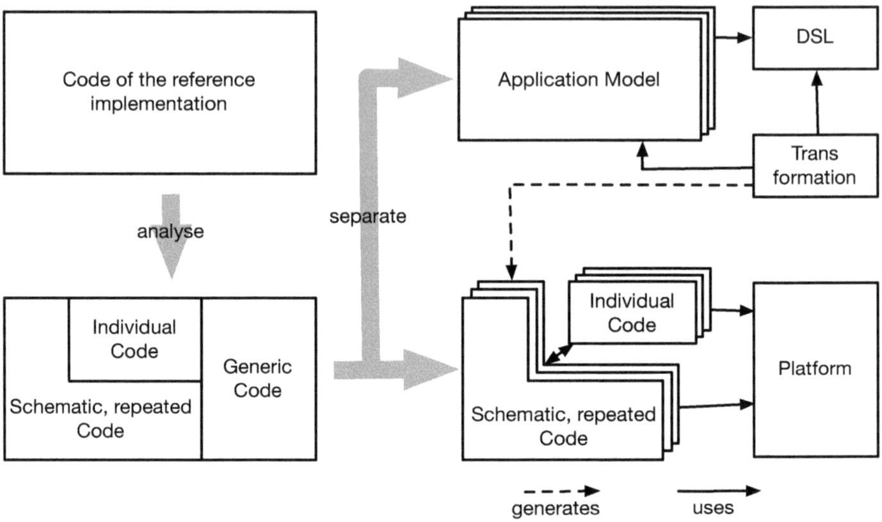

Figure 7.16: Concepts of model-driven software development [SV06]

We already discussed our motivation for the use of the Test Parameter object. Accordingly, we argue that the direct generation of program code from feature models is not well-suited for MTCC because the feature models used in the test engineering phase do not represent implementation artifacts on the level of components or subsystems but rather represent the parameterization of a single Test Step.

Figure 7.17 illustrates the concepts and the process of the *Templates and Filtering* code generation approach. The filtering phase mostly corresponds to the transformation of Test Configurations in Test Parameter objects. In the template phase, a template of a test class is used to generate program code for calling the methods in the Test Adapter that contain the implementation of a given Test Step instance.

Every method call is preceded by code that initializes the TestParamter-object that represents the configuration of the Test Step. Section 10.5 gives a concrete example of test code generation in MTCC based on the prototype implementation of MTCC used in the evaluation. Figure 10.1 on page 176 illustrates the main part of the MTCC implementation.

Figure 7.15 gives an overview of of the artifacts used in code generation. Figure 10.1 in Section 10.2 illustrates the architecture of the MTCC implementation used in the work.

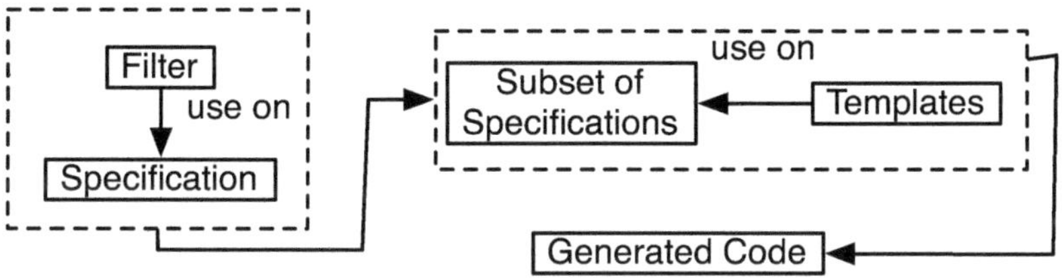

Figure 7.17: Concepts of the Template and Filtering approaches [SV06]

7.5 Chapter Summary

In order to allow the construction of tests based on the information contained in the Application Test Model that represents a SUT, the model has to be represented to domain experts in a comprehensible way. Test Configurations have to be instantiated that represent the Test Steps taken on the testee and their parameterization.

The editor represents the available Test Step instances contained in the Application Test Model using GuiBuilder classes that map concepts from the feature models to generic or domain specific GUI representations. Interactions with these GUI representations are expressed as specialization steps for the feature model. MTCC expresses specialization steps as configuration nodes in the feature model to facilitate the editing of Test Configurations without directly affecting the structure of the underlying feature model.

A central aim of MTCC is the reuse of tests, represented by Test Configurations for different SUTs. MTCC supports the reuse of tests for a system if the sequence of Test Steps defined for the test can be expressed on this system and if each Test Step supports the configuration of the Test Step in the original test.

MTCC uses code generation techniques to generate test code from Test Configuration. Tests are generated in a Templates-and-Filters approach. Generated tests utilize a Test Adapters implementation that serve as a platform that encapsulates implementation specifics necessary to execute a test on a SUT.

8 Application of MTCC to the Digital Library Domain

This chapter describes the application of MTCC for system families from the Digital Libraries domain.

Section 8.1 introduces the system family of Digital Libraries that MTCC is applied to in this work. We examine a system family of three Digital Libraries, the scientific portal Sowiport, the interdisciplinary information Service infoconnex and the IREON portal of the German Institute for International and Security Affairs, addressing international relations and area studies. In addition to the Digital Libraries, we investigate the Apache search server Solr.

Section 8.2 discusses the test-relevant Services for this system domain.

Beside the Test Steps that are specific to the requirements for a system family, MTCC also defines generic Test Steps, for instance for the comparison of variables that hold aspects of the state of a system family. These Test Steps are mostly independent from the systems, system family or application domain to which MTCC is applied. We examine these Test Steps in Section 8.3.

In Section 8.4, we discuss the system family-specific MTCC Services for the Digital Libraries considered in this work.

8.1 A System Family of Digital Libraries

The Digital Libraries Sowiport[1] and infoconnex[2] of GESIS as well as the IREON[3] portal of the German Institute for International and Security Affairs form the system family [Par76] of testees to which MTCC is applied in this work.

MTCC regards the examined systems as Digital Libraries and therefore, as discussed in Chapter 4, as Information Retrieval systems. For the purpose of this work, we limit our analysis of Digital Libraries to the search process and the presentation of results. In addition to the digital libraries, MTCC is applied to the search server Solr[4]. We examine Solr in order to assess the applicability of MTCC

[1] http://www.sowiport.de

[2] http://www.infoconnex.de

[3] http://www.ireon-portal.de

[4] http://lucene.apache.org/solr/

to systems that are not members of the system family but rather part of their underlying infrastructure.

The three Digital Libraries considered have been designed and implemented at the German Social Science Infrastructure Services (GESIS) in cooperation with external partners. Figure 8.1 illustrates the relations of the systems as well as their commonalities.

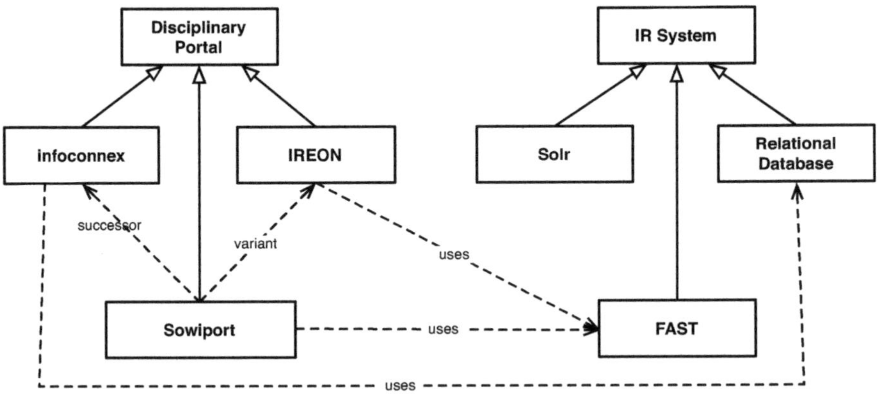

Figure 8.1: Relations of systems within system families

The central use case that is relevant for all three Digital Libraries as well as Solr is search. A user formulates a query and submits it to the system, the system generates one or more lists of documents that are likely to be relevant to the query. The user surveys the documents in the result lists, selects documents according to her or his information need and further refines the query.

Sowiport, infoconnex and IREON differ in the functionality available for searching as well as in the types of interactions that are possible with an initial list of results. Solr does not support functionality for the refinement of queries.

All considered systems are web applications. A user interacts with such systems using a web browser and the HTTP protocol.

8.1.1 Sowiport

Sowiport is a disciplinary information portal for the social sciences. It integrates 15 databases and contains 2.5 million documents in total. The system is implemented in PHP and Java; a commercial IR system is used as its search back-end. Figure 8.2 and Figure 8.3 display the simple and the advanced search form of

Sowiport. The simple search form consists of seven text input fields that support queries for specific fields of documents. By using the Check Box *Alle Wörter*, the conjunction of the terms within one text entry can be switched between Boolean AND and OR. By selecting the Check Box in the lower part of the search form, the search can be restricted to documents that were published within the last four years.

The advanced search form displayed in Figure 8.2 extends the simple search by filters that restrict a search to a subset of all available documents and by providing support for Boolean operators that allow the formulation of more complex queries. The filter displayed in the figure supports the restriction of result documents by their information type. Other filters can be used to restrict a query to certain languages and databases.

The entry fields of the advanced search form allow the explicit selection of the fields that are to be searched as well as the choice of the Boolean operator that determines the relationship of the multiple fields in search.

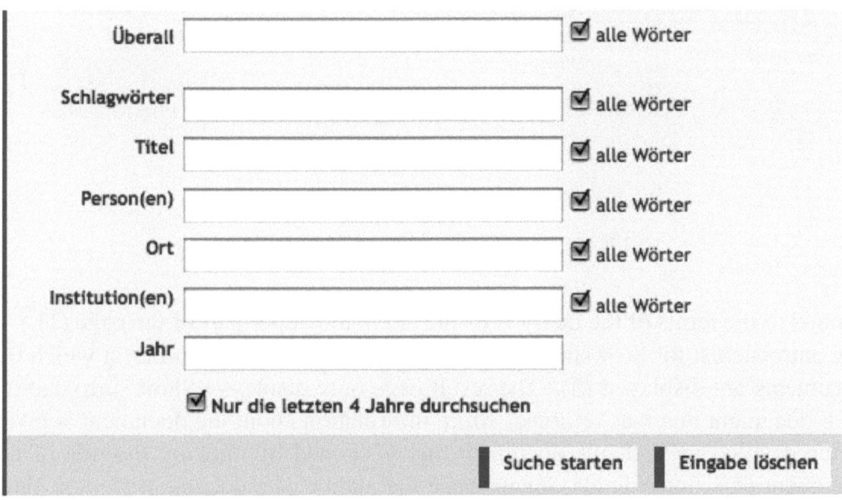

Figure 8.2: Simple search in Sowiport

Figure 8.4 displays the result page that presents the documents that are returned for a query. The result page also includes Services to refine the query. A list with short representations of the documents that were found for a query is in the center of the result page. A representation of the query with all transformations that were

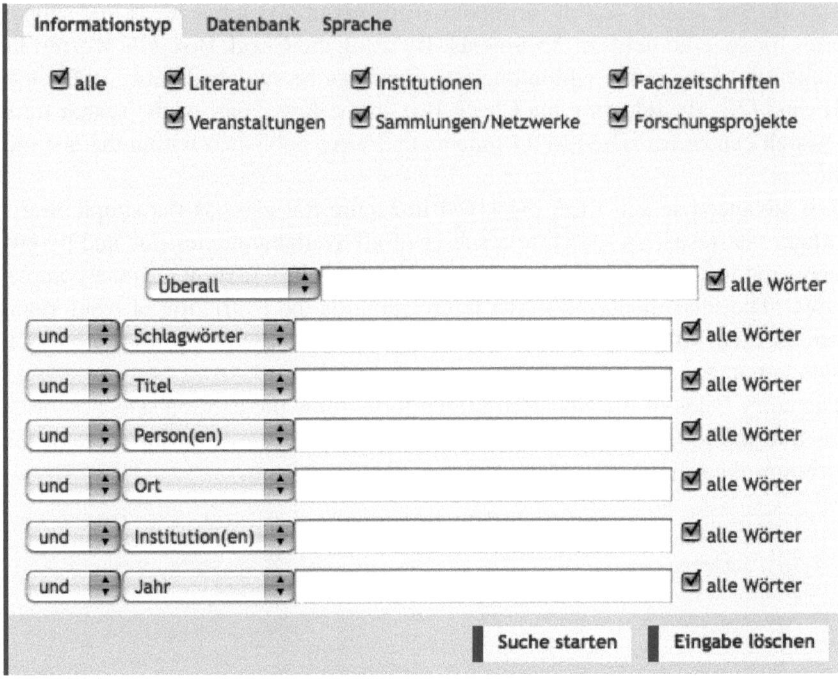

Figure 8.3: Advanced Search in Sowiport

applied to the terms of the query is displayed in the upper part of the page (1).

Controls exist for browsing the list (2) and for changing the order in which the documents are displayed (3). The result page only displays a short surrogate for each document that was returned. More information about the document is given in the detailed view of the document that is opened by clicking the title of the document (4). Some fields, for instance the author of a document (5), are links that send a new query to the Digital Library. In this case, a search for the author would be started. Other links allow users to work with reference management systems[5] or help to check whether the full text of a document is available (6). By using a Check Box (7), it is possible to mark documents for export. The links in the right part of the page are navigators. Navigators allow the restriction of a result set by facets [BYRN99] and thus the refinement of the search. Every navigator

[5] http://www.refworks.com

displays the most frequently occurring values for a given field in the result set and their count. By clicking the link associated with a value, the result set of the query is restricted to all documents that contain this value for this field.

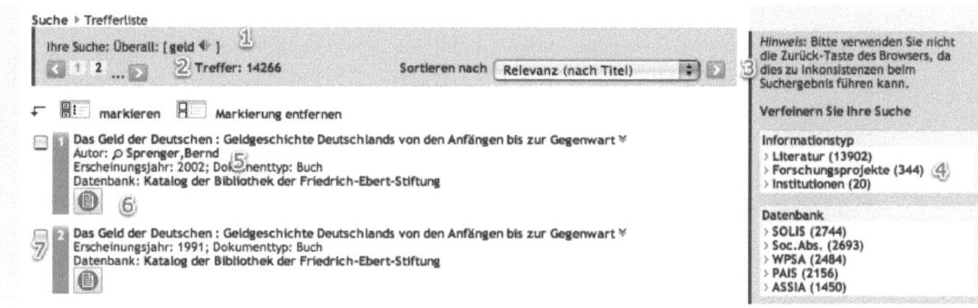

Figure 8.4: Result page in Sowiport

8.1.2 infoconnex

infoconnex is a portal that allows the integrated search of documents from the disciplines of social sciences, pedagogics and psychology. The portal is implemented in Java; the Information Retrieval component is based on a relational database.

Regarding the functionality and design of its search forms, infoconnex is similar to the simple and advanced search in Sowiport. Differences between Sowiport and infoconnex exist in the display of search results as displayed in Figure 8.5. infoconnex returns up to three result lists for a query. Each result list contains documents from one database. The currently displayed list can be selected by clicking the tab at the top of the list.

In contrast to the other Digital Libraries, infoconnex does not support navigators or the ordering of documents by relevance based on the similarity of a query and a document. infoconnex does provide a relevance ranking based on the centrality of an author for documents from the social sciences [Mut03].

8.1.3 IREON

The IREON portal provides information about international relations and area studies. The portal shares significant parts of its implementation with Sowiport, especially concerning the transformation of queries and underlying retrieval system. Sowiport and IREON also have some databases in common.

Figure 8.6 displays the advanced search for all databases in the IREON Portal.

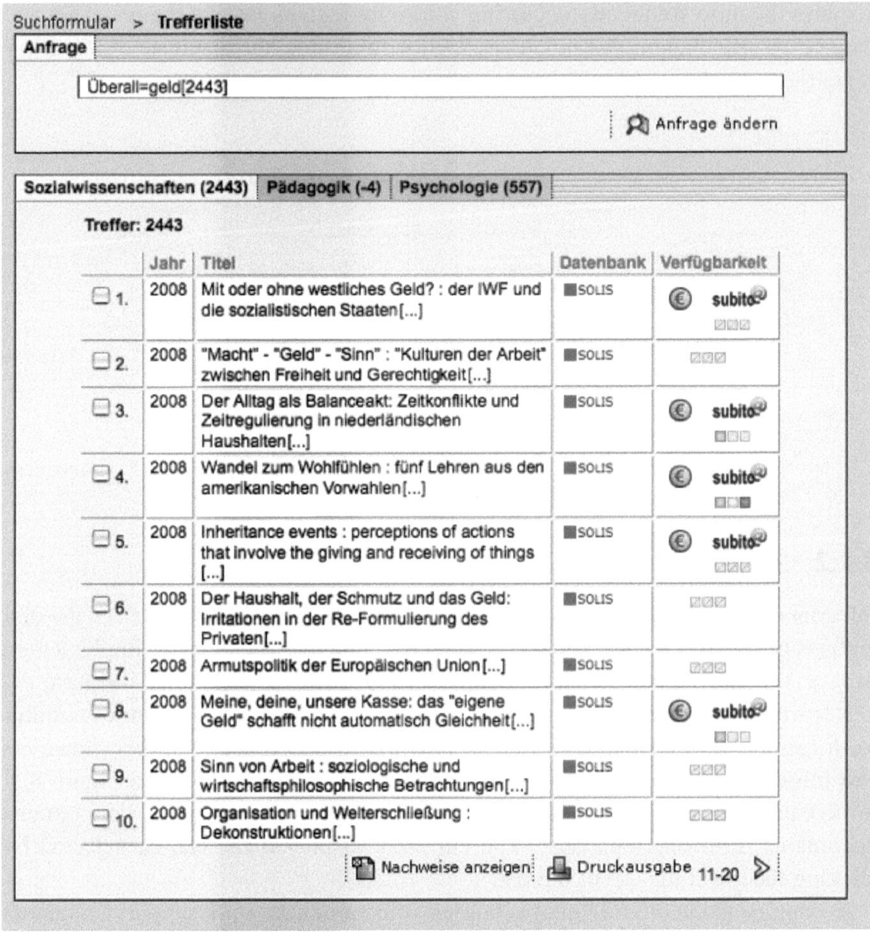

Figure 8.5: Result page in infoconnex

The functionality of search forms and filters is analogous to those in SOWIPORT. Besides the search forms displayed, IREON also provides a number of of additional search forms that are adapted for special subsets of the portal.

Figure 8.6 displays the result page in IREON. The functionality of the elements in the user interface is similar to Sowiport.

Figure 8.6: Advanced search in IREON

8.1.4 GESIS Solr Installation

In contrast to the Digital Libraries introduced so far, the Solr installation of the GESIS is a search server, not a Digital Library. The Solr search server only offers a minimal HTML-based user interface that supports the submitting of queries to a system and provides statistics about the operations of the system. The GESIS Solr installation includes the same document collections as the Sowiport and IREON portals. It is used for automated search tasks and complex queries for which the features of a Digital Library are not needed.

We include the Solr installation at the GESIS in our discussion since it represents a part of the low-level infrastructure that would be used by a Digital Library. By including Solr, we investigate whether MTCC can be used for the testing of such sub systems of Digital Libraries as well as for testing the libraries themselves.

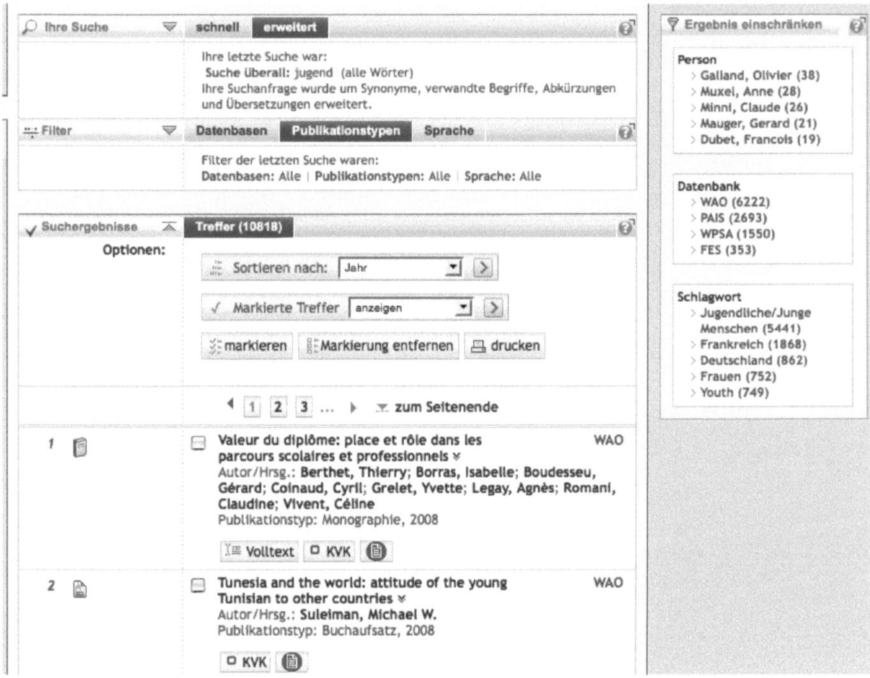

Figure 8.7: Result page in IREON

8.2 Test-Relevant Services and Test Steps

In order to test the systems introduced in Section 8.1, they have to be described in terms of their Services and Test Steps. The Services and Test Steps relevant for the system family considered in this work are presented in Sections 8.3 and 8.4.

8.2.1 Manipulation of the SUT

We identify the test-relevant Services and Test Steps for the systems of the considered system family by examining the test sets discussed in Section 9.3.1. We argue that these Test Steps represent both current testing practice and the requirements of the domain experts to a testing system.

Each test set holds a number of tests or documented faults that can serve as the base of a test for one of the systems in the system family. The different test sets come from sources such as previously defined manual and automated tests. Manual tests are specified by domain experts, and documented, testable faults for a system.

Based on the test sets, the most important Services of a Digital Library are the search for documents and the display of those documents that were returned as the result of a search. In addition to these core functionalities, the various options that allow the refinements of queries are also of importance. MTCC must include models for Services and Test Steps that represent these functionalities and their interactions with a user. An additional requirement for MTCC is the ability to model retrieval tests.

An application of MTCC for Digital Libraries must include models that represent the functionality to formulate a query and submit it to the system. This model of the search Service must represent all options that the system provides for searching. Furthermore, MTCC must provide Services that represent a list of potentially relevant documents returned in the result page as well as the functionality provided to further refine this list or otherwise interact with it. MTCC must also include Test Steps that represent the interactions performed with a Service for example the entry of search terms into a search form, the selection of filter values or the interactions with document surrogates from a list of results.

The Services and Test Steps that support the refinement of a search are an important aspect of the Digital Libraries discussed here. They address the problem discussed in Chapter 4 that a user will often start a search with only a vague definition of his or her information need but will gradually refine it in the process of a search. Among the Services that serve this purpose are the following:

- The navigators in Sowiport and IREON support the restriction of a result set of documents to those documents that contain a specific value for a field.

- The order, in which the documents of a result set are presented to the user can be adapted to the preferences of the user. The systems discussed here support the arrangement of documents by different concepts of relevance or the sorting of documents based on the contents of fields.

- The link search in IREON and Sowiport allows the start of a new search based on one of the results of the previous search. For example, a click on the author of a document starts a search for all documents by this author.

MTCC represents each function above as an independent Service We argue that such a representation is advantageous compared to the extension of existing Services: The complexity of the individual Services is limited, making maintenance easier; and functionality can be added to the models of a system without changing existing Service An advantage of the MTCC modeling approach is that fine-grained Service models allow for a more precise description of the behavior of a system in the Application State Model.

8.2.2 Verification of the Testee

In order to test a system it is not only necessary to exercise the system and interact with its provided functionality, but it is also necessary to verify that the behavior of the system and its visible states conform to the expectations expressed in a test. The Services and Test Steps provided in MTCC for the verification of a system fall into two groups, those used to capture some visible aspect of the system state in a variable and those that compare the values of these variables with the expectations of the test:

- The aspects of the system states of a Digital Library that MTCC must be able to capture and store, include the number of documents in a result set, the fields included in the documents and the values of these fields. This also includes the values and counts with navigators. MTCC represents the inspection of the system state by Test Steps of the type assertion, meaning that the state of the testee itself is not manipulated, only captured. MTCC uses variables to store the captured aspect of the system state.

- Test Steps that verify the state of a system operate on the variables. Test Steps of this kind compare numeric or textual values, test the values in two lists for specific subset or superset relationships and determine Information Retrieval metrics like recall and precision.

As described above, MTCC uses variables to store an aspect of the system state for later comparison. The use of variables allows to separate Test Steps used for capturing the system state from those that test the system state against a condition. This separation helps to reduce the complexity of individual Test Steps and lowers the number of Test Steps that must be implemented for a system.

Variables are always expressed as lists of values. Single values are represented by a list with a single element. MTCC does not implement a type system, every entry in a variable list is implemented as a string. The interpretation of this string depends on the Test Step that uses a variable. Depending on the Test Step, textual or numeric interpretations are possible.

8.3 Generic Services

Generic Services and the Test Steps that refer to these Services are mostly independent from the system that is tested by MTCC.

Two types of generic Services exist in MTCC, system-independent Services and domain-independent Services. System-independent Services represent characteristic functionality for the application domain that MTCC is applied to, but Service instances do not differ between systems. The Test Steps of domain-independent Services represent functionality that is needed for testing with MTCC, but that is not specific for a particular system family or system. Most domain-independent Test Steps are assigned to the DEFAULT_SERVICE.

8.3.1 System-Independent Services

The feature model of system-independent Services is identical for every instance in every Context and in every system. Since every instance of a Service is identical to every other instance, the Service instances do not contain any information and serve purely as markers for the availability of the Service in a particular Context.

The Service LINK marks Contexts that support the change to another Context without configuration of a Test Step interface. For the system family of web-based Digital Libraries considered in this work, the link Service represents the HTML links between different pages of the digital library. Information about the available targets of a LINK Service instance is included in the Application State Model and in the dynamic part of the Application Test Model. The feature model of an instance of the LINK Service only marks the existence of one or more links in a system. The Test Step FOLLOW_LINK represents the following of a link. Since the activation of a link changes the current Context of the system, this Test Step is an action.

The Service EXTERNAL_TEXT is used for Contexts where test-relevant functionality is not or not completely represented by MTCC Service instances. It is used in cases where Contexts have to be tested that are not part of the system modeled with MTCC. For web applications, one example for such Contexts would be external pages, reachable by links from the modeled system. The Test Step SAVE_EXTERNAL_TEXT is a SameStateAction that stores all text that is displayed in a Context, for example, all textual content displayed on a web page.

8.3.2 Domain-Independent Services

Domain-independent Services represent functionality that is specific to MTCC and mostly independent from the system or system family under test. The implementation of MTCC examined in this work only defines one domain-independent Service the DEFAULT_SERVICE. This Service is instantiated for every Context of a system. A number of Test Steps are assigned to the DEFAULT_SERVICE. All of these Test Steps with the exception of the RESET_SYSTEM Service are assertions, they do not change the current Context of the system or manipulate its state:

- The Test Step SET_VARIABLE assigns a user defined value to a variable. This variable can later be used in other Test Steps, for example, in comparisons. Two variants of this Test Step exist.
  The Test Step SET_VARIABLE_LONG_TEXT is represented in the editor by a multi-line text input field and allows the entry of long texts. The Test Step SET_VARIABLE_FROM_FILE is used to store values read from a file into a variable.
- The Test Steps COMPARE_TWO_VARIABLES and CHECK_VARIABLE are used to compare variables with expected values The Test Steps CHECK_VARIABLE requires the value to be compared against a value to be entered in the editor while COMPARE_TWO_VARIABLES is used to compare two previously defined variables.

  Since MTCC does not use a type system for the values in the variables, both Test Steps require the user to specify the data type used for the comparison. In addition, the user must select whether the values in the variables are expected to be equal, unequal or of another relation.

  The Test Step COMPARE_TWO_VARIABLE_LISTS is used to compare two variables that each hold multiple values. The values in the variables can either be treated as sets or ordered lists. They can be tested for equality and for is-part-of relationships.
- The Test Step CHECK_SORTING_OF_VARIABLE checks whether the values in a variable are sorted. Values can be interpreted as numbers or as strings.
- The Test Step RESET_SYSTEM is an action that return the system to its initial state before any Test Steps were executed on it. How this Test

Step is implemented depends on the SUT. In the case of the digital libraries discussed here, the Test Step resets the session state of the SUT, emulating a user logging out.

- The Test Step COUNT_DISTINCT_VALUES_IN_VARIABLE calculates the cardinal number of the set of values in the variable and saves the result in another variable.

8.4 Domain-Specific Services

In the following we discuss the system family specific Services that represent the functionality of the Digital Libraries discussed in Section 8.1. For each Service we discuss its assigned Test Steps. The Services and Test Steps were defined based on an analysis of the concepts required for the modeling of the test sets defined in Section 9.3.1

Every Service discussed in the following is represented by a feature model. All feature models have the following two feature nodes in common: The Type node that represents the class of the Service and the Name node that contains the unique name of this particular Service instance.

The feature models represents Services as defined in the Domain Feature Model. A Service instance would be a specialization of this model. We discuss both the system family and system-level models for complex Services.

8.4.1 The SEARCH Service

The SEARCH Service represented as a feature model in Figure 8.8, represents a search form. A search form in the model consists of multiple input fields, represented by an InputField node. For search forms that allow the selection of the queried field for each input field, the feature group FieldNames represents all selectable fields for a single InputField. The feature groups TermRelations and FieldRelations represent the relations of multiple terms within a single input field and of multiple input fields to each other. The FieldValue attribute is configured in test modeling and specifies the subject of the query for this input field.

The filters that are available for restricting a search are modeled as a FilterForm. Every FilterField represents a single filter with multiple selectable values. The feature FilterName describes the field that the filter refers to, for example, the language or the document type of a bibliographic record. Possible values for the filter are defined in the feature group FieldValues.

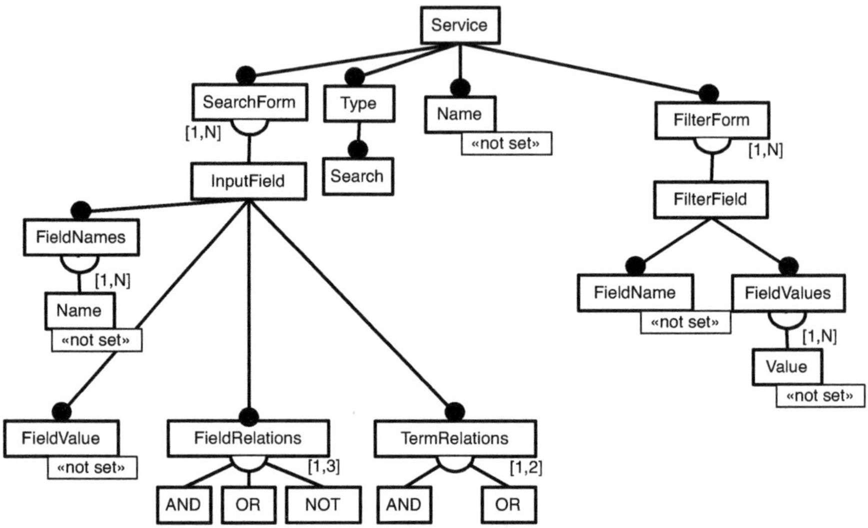

Figure 8.8: Feature model of the SEARCH Service at the domain level

Figure 8.9 displays the feature model of a Service instance. In the Service instance, it is possible to search either in the field title or in the field person. The different input fields represented by the model are always in an AND relationship, the terms in an input field can be either in an AND or OR relation.

Two Test Steps are defined for the SEARCH Service One Test Step is used to submit a search, the other to set the filters that limit a search to a subset of all documents in a database. The Test Step START_SEARCH is an action that represents the submission of a search. The Test Step SET_FILTER is used for the selection of filter values. Since the Test Step does not cause a change of Context, it is a SameStateAction.

8.4.2 The SEARCH_OPTIONS Service

The SEARCH_OPTIONS Service displayed in Figure 8.10, represents options that are available to influence the processing of queries. The only search option available for the Digital Libraries considered here are cross-concordances. The SEARCH_OPTIONS Service is a separate Service from the SEARCH Service which allows the representation of a system where search options are set globally for all search forms.

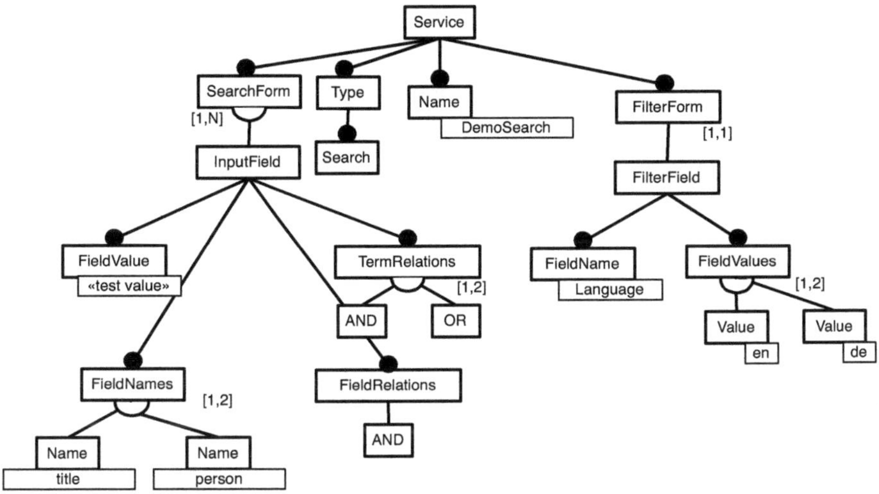

Figure 8.9: Instance of the SEARCH Service represented as a feature model

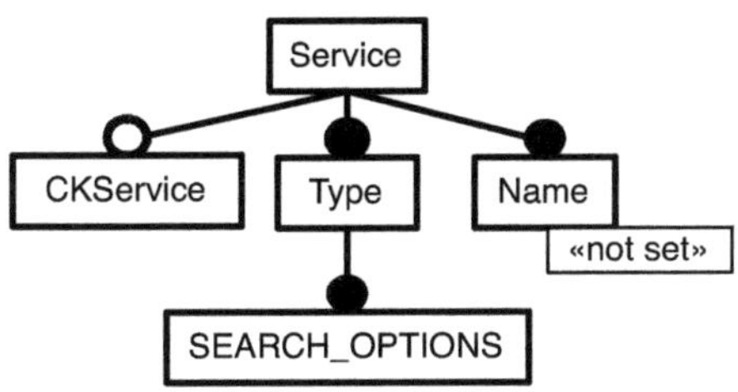

Figure 8.10: Feature model of the SEARCH_OPTIONS service

8.4.3 The REQUEST_INFORMATION Service

The Service REQUEST_INFORMATION, represented in Figure 8.11, models the presentation and explanation of a query after it was executed by the system. In the Digital Libraries considered here, this Service informs about the transformations that were applied to a query during query expansion. The Test Step SAVE_QUERY_INFORMATION is assigned to the Service This Test Step stores the information displayed by the Service in a variable.

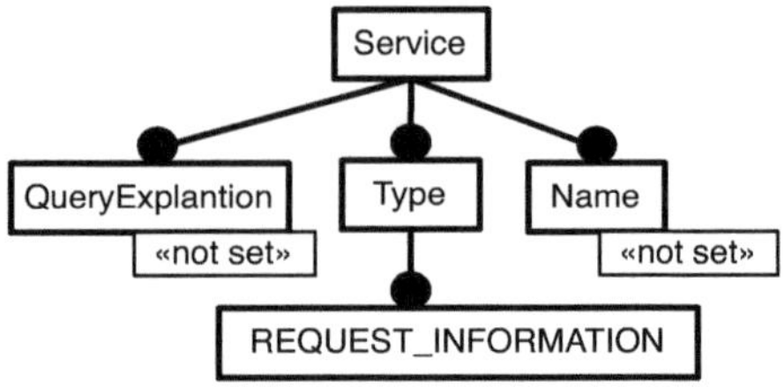

Figure 8.11: Feature model of the REQUEST_INFORMATION Service

8.4.4 The DOCUMENT_LIST Service

The DOCUMENT_LIST Service represents a list of documents returned for a query. In the feature model in Figure 8.12, this list is characterized by three feature nodes: The fields that each of the result documents can potentially contain are represented by the FieldNames node. This feature determines which fields of result documents can be checked against expectations and stored in variables.

The number of documents that can be simultaneously displayed is given by the feature PageLength. The SupportsSelection node reflects whether it is possible to mark documents, for example, in order to export them.

The DOCUMENT_LIST is not only used to present an overview of the results of a query, but also for the detailed display of documents. The difference between a DOCUMENT_LIST instance used to represent an overview of documents on the result page and an instance that represents one ore more documents in detail is that the latter has a PageLength of one and contains all FieldNames in the result overview. The feature GroupsDuplicates indicates if the result list supports the marking of duplicates in the result set.

The DOCUMENT_LIST Service has the greatest number of assigned Test Steps of all Services. This reflects the central role that the completeness and correctness of search results has in the Digital Library domain:

- The SameStateAction SAVE_NUMBER_OF_RESULTS stores the absolute number of documents that were returned as the result of a query.
- The SameStateAction SAVE_FIELD_FROM_DOCUMENT stores the value of one field for one document in a variable.
- The SameStateAction SAVE_FIELD_FOR_ALL_RESULTS stores the values of one field for all documents within a range of the result list.

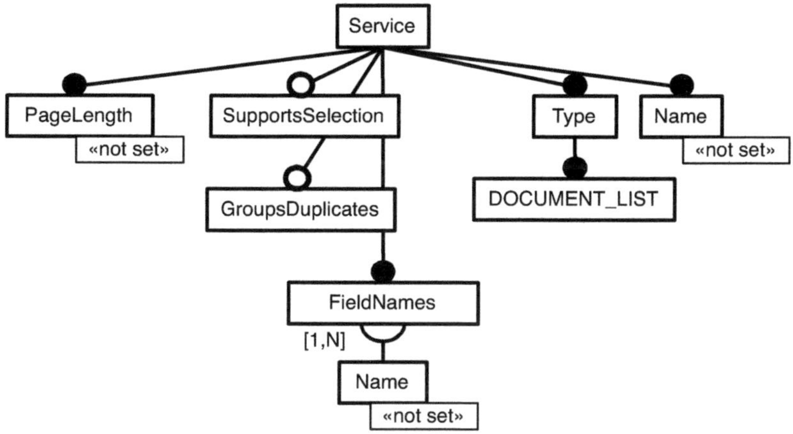

Figure 8.12: Feature model of the DOCUMENT_LIST service

- The Assertion Test Step CHECK_DUPLICATE_GROUPING verifies if two documents are marked as duplicates.
- The SameStateAction MARK_DOCUMENT marks a document on the current page of the result list.
- The Test Step CHECK_RECALL_AND_PRECISION_RESULTS is used to calculate recall and precision for a section of the result list and to store the result in a variable. A prerequisite for the invocation of this Test Step is that the IDs of all documents are available from the list of documents and that relevance assessments are available either for all documents or for a pool of documents.

8.4.5 The MARKED_DOCUMENT_HANDLER Service

The Service MARKED_DOCUMENT_HANDLER, displayed in Figure 8.13 represents the invocation of an action on some previously selected documents. Documents are selected with the Test Step MARK_DOCUMENT. An example for such an action is an export in the BibTex format. All available actions are described by the feature group ExportActions. Only the Action Test Step EXPORT_MARKED_DOCUMENT is assigned to the Service This Test Step allows the selection and execution of an action that is to be invoked on the documents by its name.

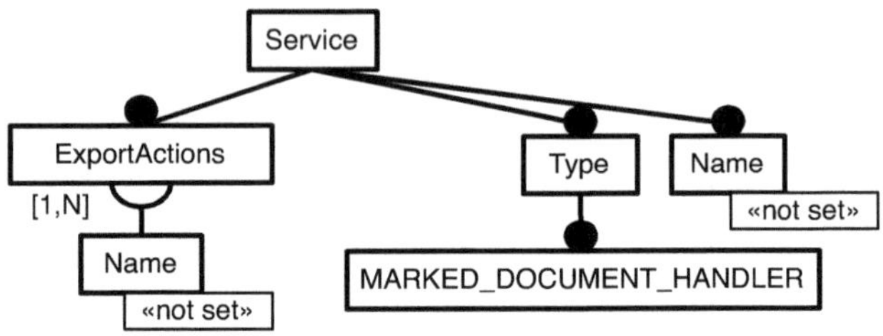

Figure 8.13: Feature model of the MARKED_DOCUMENT_HANDLER service

8.4.6 The ORDERING Service

The ordering Service describes the different criteria by which the documents in the result list can be ordered. Criteria include the fields by which the result list can be sorted, as well different concepts of relevance or ranking. The feature group OrderFields in Figure 8.14 specifies for each criterion whether it supports ascending and/or descending order. The Name attribute contains the name of the criterion. The Action CHANGE_RESULTLIST_SORTING is used to invoke the ORDERING Service

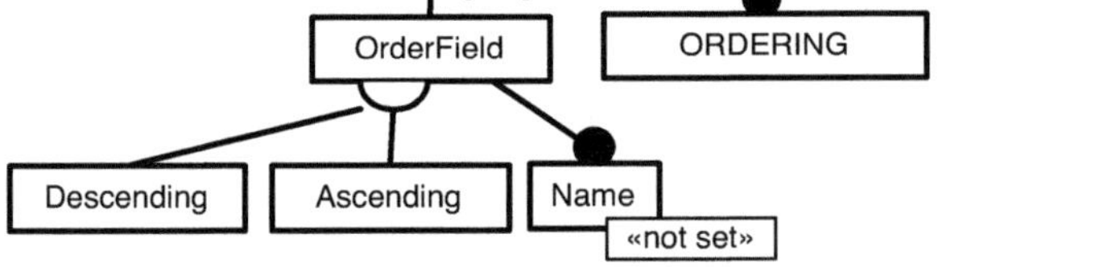

Figure 8.14: Feature model of the ORDERING service

8.4.7 The DOCUMENT_LINKS Service

The DOCUMENT_LINKS Service represents the functionality to start a *link search* based on values of a field in the result documents. In a link search, the search term is not entered by the user. Instead, the values for some fields of documents are represented as links. Following such a link starts a new search for the value of the link. Sowiport, for instance, supports a link search for all publications by an author by clicking on the author's name. The FieldNames feature group describes the fields of result documents that support a link search.

The Action Test Step FOLLOW_LINK_IN_DOCUMENT is assigned to the Service DOCUMENT_LINKS. This Test Step allows the selection of documents and fields within documents to be used for a link search.

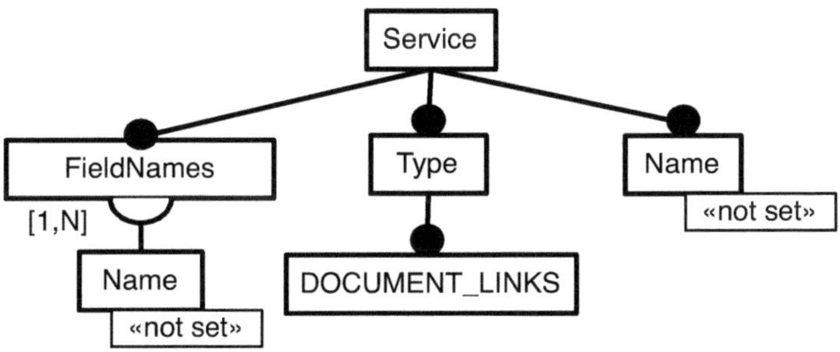

Figure 8.15: Feature model of the DOCUMENT_LINKS Service

8.4.8 The PAGINATION Service

The PAGINATION Service describes the different ways in which a user can browse the different pages of a result list for a search. Figure 8.16 displays the feature model of the Service The features Absolute and Relative describe whether jumps over sections of the result list are possible The Test Step GO_TO_PAGE_IN_LIST, a SameStateAction, allows the change of the current page in the list of results.

8.4.9 The NAVIGATOR_LIST Service

The Service NAVIGATOR_LIST represents the navigators that allow the restriction of a result list to documents that have specific values for certain fields. Fig-

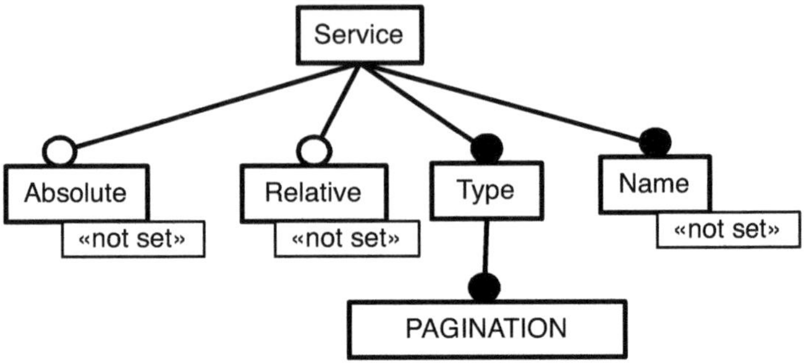

Figure 8.16: Feature model of the PAGINATION Service

ure 8.17 illustrates the feature model of the Service The CompleteDisplay feature describes whether the navigator returns all values for a facet or, as is the case for the Digital Libraries considered here, only the most frequent values.

The NavigatorField node contains the name of the field that is represented by the navigator. The feature group Values includes the possible values for facets. Three Test Steps are assigned to the Service the SameStateActions SAVE_NAVIGATOR_COUNT, and SAVE_NAVIGATOR_VALUES, and the Action FOLLOW_NAVIGATOR_LINK:

- The Test Step SAVE_NAVIGATOR_COUNT stores the number of results for one value of a facet.
- The Test Step SAVE_NAVIGATOR_VALUES saves all values of a Navigator for a field in a variable.
- The FOLLOW_NAVIGATOR_LINK represents the selection of a navigator value and the restriction of the result list to documents that contain this value for the field. The Test Step is an Action.

8.5 Chapter Summary

MTCC is applied to a system family of three Digital Libraries and an installation of the Solr search server. These systems are analyzed for their test-relevant Services and the Test Steps that exercise these Services. Services are test-relevant if they represent features or behaviors of the SUT that is relevant for the domain experts to

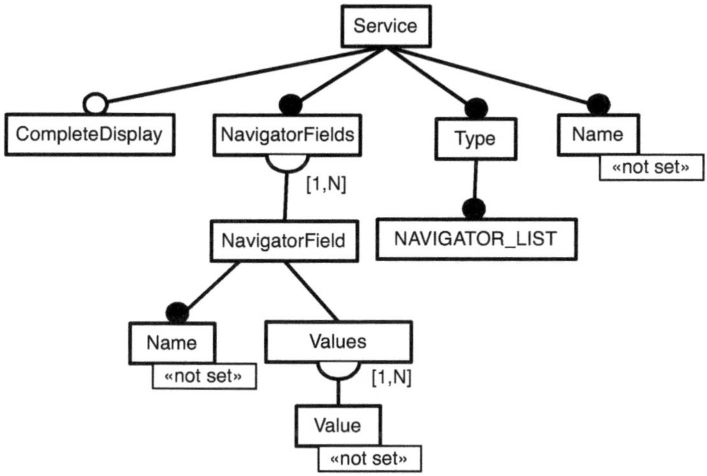

Figure 8.17: Feature model of the NAVIGATOR_LIST Service

assess the quality of a system. MTCC distinguishes between domain-specific and generic Services. Domain-specific Services represent functionality that is characteristic for the system family to which MTCC is applied. Generic Services represent functionality of MTCC, for example the use of variables to store and verify aspects of the state of the SUT.

Test Steps are test-relevant if they either manipulate the state of a Service or serve to verify that the SUT is in a certain state. The Services for the system family of Digital Libraries considered in this work focus on the functionality of these system for searching and displaying results.

9 Validation Goals and Design

In this chapter, we discuss the role of validation in software engineering. We apply the GQM approach to the validation of MTCC and formulate the goals and questions based on which MTCC is evaluated.

Section 9.1 examines the role of validation in software engineering. The need for a validation is motivated and different approaches to the validation of software are discussed.

Section 9.2 defines the goals of the MTCC validation. We investigate how different types of validation apply to MTCC.

Section 9.3 discusses the application of the GQM approach to the validation of MTCC in detail. Five validation questions are formulated in order to address the general feasibility of the approach as well as its practicability.

9.1 Validation in Software Engineering

A systematic validation is a prerequisite for proving the effectiveness and efficiency of a specific method or technology and for the verification of a hypothesis [Pre01, WRH$^+$00]. How a validation is conducted is determined by the available resources [Pre01].

In this work, we employ the GQM approach in the context of a case study to validate both the feasibility of the MTCC and its practicality. The validation of the efficiency of the approach is the subject of future work.

9.1.1 Reasons for Validation

In software engineering, the validation of assumptions about the effectiveness and efficiency of a method or tool is necessary for the same reasons that the validation of hypotheses is necessary in science. Specifically, validation has two goals:

- The validation process assesses the general feasibility, practicality and efficiency of the object of the validation. Such an object might be a tool, a process or an algorithm. Here, the validation provides the basis for the future improvement of the object of validation [Bas96].

- The discipline of software engineering in its entirety benefits from the experience of the conducted validations of past approaches. The experience and knowledge gained serve as the basis for further research.

Despite these facts, research in software engineering often ignores validation. The reasons come in part from the tradition of software engineering. Additionally, a number of practical reasons that are specific to software engineering hinder the conducting of robust validations for the discipline [PPV00].

In practice, the optimal validation of an approach or tool will not be possible in most cases since the resources necessary to conduct a perfect validation are prohibitively high [HHK^{+}05, Tic00]. We argue therefore that requirements for validation approaches must take the available resources into account. Based on this argument we adopt the three types of validation defined by Freiling [FER08]. They address the feasibility, practicality and efficiency of an approach. The separate validation of the different properties allows the successive evaluation of a system and the evaluation of fundamental properties even with limited resources.

If software engineering is considered in the short term, then the purpose of the discipline is to provide methods that support the creation of high quality systems (see Section 1.1 for the meaning of high quality in this context). Considered for the long term, the objective of software engineering is the creation of theories that support a better understanding of the actors, processes and artifacts relevant to the field. These long-term and short-term views correspond to the roles of the practitioner and the researcher [Bas96]. Validation is an essential means to reach both the short term and long term goals of software engineering.

One aspect of validation is the assessment of a product or process with the goal to raise the quality of the system. This aspect is addressed by a number of approaches from manufacturing and quality management, especially methods like Continuous Improvement [CAJ01]. One concrete tool to assess the consequences of changes is the Plan-Do-Check-Act cycle, also known as the Deming Wheel [CAJ01]. The Plan phase of the approach develops hypotheses that are put into practice in the Do phase. The Check phase measures the effects of the changes; the Act phase makes changes to the production process based on the experiences of the proceeding phases and develops new hypotheses.

In contrast to manufacturing, the product of a software development process will usually only be created once [Bas96]. However, since software can be developed in increments or iterations, experience from one iteration can be used to make improvements to the process.

Software systems are defined by a large number of quality attributes (see Section 1.1). This leads to a situation where the development of measures that support an assessment of a software system is non-trivial. If the requirements for a system are known and the degree, to which these requirements are met can therefore be measured, empirical methods [ZW98] and the definition of the appropriate metrics [Kan02] are essential means to assess trends in quality, thereby facilitating the design and implementation of higher quality systems.

The validation of assumptions is an essential aspect of the empirical method. Every hypothesis has to be provable or falsifiable by repeated verification [PPV00]. For software engineering as an engineering science discipline it follows that a hypothesis or the predictions made for a process or software system must be compared with the behavior of the system. Empirical Software Engineering propagates the use of empirical methods in software engineering [TLPH95, PPV00]. Such methods are also the basis of physics [Bas96] or evidence-based medicine [KDJ04].

9.1.2 Challenges to Validation

A number of reasons exist for the relatively low adoption of empirical methods in software engineering. One aspect that hinders the application of such methods is the problem of comparability and reproducibility of results. Another factor is lack of knowledge about the required methods.

One central cause for the lack of reproducibility and for the difficulty in comparing different results is the high degree to which the software development process depends on the skills and abilities of the participants and their individual approach to software development [Bas96]. Beside differences in the participants, other factors also have a negative impact on comparability and reproducibility. These factors include differences in the software development process, the programming languages and tools used and the specifics of the organization in which a software system is developed.

A controlled experiment limits the influence of these external factors [WRH+00], but in order to assess the consequences of an approach in practice, it is desirable to evaluate the approach in the field. An important factor in this context is that the control over the evaluation conditions is significantly lower in the field than in a controlled experiment.

In order to apply empirical methods to software engineering, it is necessary to involve human participants. Such involvement results in a considerable increase in the resources necessary to conduct the validation. Furthermore, the resources increase with the level of control that is needed over the validation setting. As a consequence of the above, only a small fraction of the validations done in software engineering are controlled experiments [HHK+05].

While the awareness of the need for empirical evaluation is rising, knowledge of the necessary methods and processes is still lacking in the community [ZW98].

Compromises about the scope and conducting of a validation cannot be avoided in practice. It is important not to accidently affect the validity of the validation. In practice, this means that only those aspects of a system can be validated for which hypotheses can be proven or falsified with the resources available for the validation.

In the following, we discuss the different types of validation and how they can be applied to MTCC with the resources available for this work.

9.1.3 Validation Methods

Different validation methods can be distinguished by the degree of control over both the object of the validation and the data acquisition process.

Following the definition by Prechelt [Pre01], we distinguish the empirical validation methods experiment, case study, interview and literature survey. Each of these methods exerts less control over the validation setting than the preceding one. The above validation methods are mostly orthogonal to the validated aspect of a method or tool. These aspects are the feasibility, the practicality and the efficiency.

A validation must be conducted based on a systematic approach. One such approach is the Goal-Question-Metric approach (GQM) [BCR94]. In the GQM, the goal for a tool or method is defined. Based on this goal, questions are formulated that assess whether a goal was reached. These questions in turn are based on metrics.

In the following we discuss the properties of the four methods of validation introduced above and examine their respective advantages and disadvantages.

- In a controlled experiment, the object of the validation and its environment are under the total control of the person conducting the experiment. Such a setting allows the exact control of the dependent variables of an experiment and therefore the precise assessment of the influence of changes on the object of the experiment. In order to limit the influence of the individual characteristics of the participants in the experiment, it is either necessary to conduct an experiment with a sufficiently large control group or to repeat the experiment with different groups [Mar07, Pre01].

  In addition to controlled experiments that are conducted in a laboratory, natural experiments are conducted in the field and investigate the object of the validation in its natural environment [Pre01].

- A case study examines the concrete application of an approach for one example [Pre01]. A case study can either be conducted in the field or under laboratory conditions. Compared to an experiment, a case study offers less control over independent variables but also requires less resources than an experiment. The definition of a case study used here differs from the view of a case study as the observation of a project outside of a laboratory [ZW98]. While the definition used above does include scenarios where a method or tool is applied with the express purpose to validate this tool, we argue that a case study differs from an *assertion* as defined by Zelkowitz. An assertion validates a method or tool solely based on criteria developed by the developer of the approach; a case study validates an approach based on external criteria.
- In a survey, a group of people is presented with a number of qualitative or quantitative questions. A survey can either be conducted as a personal interview or through a (online) questionnaire [PCFJ03]. Regardless of the modus in which it is conducted, a survey does not have any control over the object that is validated but the collected data is under the control of the experimenter.
- Literature studies and meta-analyses examine prior empirical research for the subject area. Literature surveys or literature studies are historical methods [ZW98] that neither allow control over the object of validation nor over the data that is collected. The quality of a literature study is determined by the availability and selection of relevant sources.

Of the four validation methods described above, the controlled experiment is the best in terms of control over the validation process but also needs the most resources. Based on the available resources, the validation of MTCC is therefore conducted as a case study. This approach supports a validation of the essential hypotheses underlying the MTCC approach and serves as a basis for future validations.

9.1.3.1 Validation Types

We differentiate three types of validation [FER08, Koz08, Bec08] that consider consecutively more complex and harder to validate properties of a method or tool.

A type I validation assesses the feasibility of an approach. This type of validation examines whether the approach can be implemented with limited resources. A successful type I validation is a prerequisite for the conducting of a type II validation. In the case of MTCC, that means the ability to represent the members of a system family and test for these systems as models, to transfer tests between different systems and to generate and execute tests from the models.

A type II validation considers the practicality of a method or tool. An important aspect of the practicality of an approach is the degree to which it can be applied by its intended user. The conducting of a type III validation is only possible when the practicality of an approach is proven in the type II validation. For MTCC, the type II validation considers the question whether domain experts understand the concepts and implementation of the editor and are able to use the editor for test construction.

A type III validation evaluates the efficiency of an approach in terms of the resources necessary for its application and the associated benefits. The type III validation is the most resource intensive of the three validation types. To allow a robust assessment of the efficiency of an approach, a controlled experiment that compares the approach with current practices is a desirable but costly approach.

For this work, we conduct type I and type II validations of MTCC. The conducting of a full type III validation is the object of future work — only some aspects concerning the efficiency of MTCC are evaluated.

9.1.3.2 The Goal Question Metric Approach

One prerequisite for a validation is the exact definition of the aspects of a method or tool that are assessed and the derivation of metrics used for measurement. In this work, we use the GQM approach [BCR94, BSL99] as the basis of our validation. In the GQM approach a validation is based on a goal. Questions are defined that allow the comparison of the goal and the implemented system, the questions in turn are based on metrics:

- The goal defines the relevant properties of the object under validation for the purpose of the validation. A goal is defined by the focus of the validation, the object of the validation, the issue of interest and the viewpoint from which the object is investigated.

- Multiple questions are defined for a goal, each of these questions addresses one aspect of the goal. Questions can be quantitative or qualitative in nature.

- Metrics are derived from the questions. Multiple metrics can be defined for each question. Metrics can be either objective of subjective. Objective metrics only depend on the object of the validation, subjective metrics also depend on the viewpoint of the validation.

GQM is a result-oriented approach [Zub98]. Metrics are selected for the purpose to answer specific questions about the goal of the system.

9.2 Goals and Questions of the MTCC Validation

The GQM validation of MTCC addresses five questions. The type I validation of MTCC is based on the following three questions that assess: (1) the ability of the MTCC models to represent systems and tests for the considered application domain, (2) the reusability of Test Configurations for multiple systems within the system family, and (3) the ability of MTCC to generate test cases from the models and execute them on the testee. The type II validation consists of one question. This question examines whether domain experts understand the concepts of the MTCC approach and are able to use the prototype implementation of the MTCC editor to construct tests. We only conduct a very limited type III validation in the context of this work. The corresponding question compares the resources needed for test execution with MTCC with those necessary for an existing, manual testing regime.

The basic hypothesis of this work is that the involvement of domain experts in the testing process in the context of system families is possible on the basis of models and that such a testing approach has a positive influence on the quality of a system.

Since MTCC is not a replacement of existing testing processes and since the effectiveness of MTCC in discovering faults is mostly determined by the knowledge and abilities of the domain experts that are involved in the testing process, the validation of MTCC differs from the validation of testing approaches that put less emphasis on the involvement of domain experts. As a consequence, we do not use the number of identified faults as a metric in the validation. The goal of the MTCC validation is to prove the suitability of the approach for the automated testing of Digital Libraries from the point of view of domain experts.

9.2.1 Hypothesis

The central hypothesis of the MTCC approach is that by constructing suitable models and by representing them in an editor, it is possible to support test construction by domain experts who lack formal modeling or programming skills. Furthermore, we advance the view that the abstract test models thus created can be used to test real systems and that these tests can be reused for different systems if they support the necessary features for test execution. We further argue that testing with MTCC is more efficient than manual test execution or other approaches to test automation, especially in a system family context.

9.2.2 Basic Assumptions of MTCC

One basic assumption of MTCC is that the involvement of domain experts in testing benefits the quality of a system. As discussed in Section 1.2, we argue that the knowledge of domain experts of the requirements for a system can be a form of executable specification [Fow06, Bec02, MC05b]. The usefulness of tests for quality assurance of software [Bei90, MSBT04] and the advantages of test automation [Bin99, GF00] are accepted both in the software engineering community and in practice. The involvement of domain experts or customers in the software development and testing process is well established in agile software development approaches [MC05b, And04b, AB04]. The goal to involve domain experts in the testing process is also documented outside the agile community [AOS04].

The concept of system families is long established in software engineering [Par76]. From the definition of system families in Chapter 3 follows that the members of system families are likely to share requirements and therefore also tests that verify these requirements. It further follows that the reuse of tests and test infrastructure becomes more important.

Since the members of a system family are often not systematically derived from a common set of core assets but are rather results of an opportunistic reuse process, test descriptions for the members of such a system family must be based on an abstract model of the systems, not on their implementation.

We argue that MTCC addresses both the inclusion of domain experts in the testing process and the reuse of test assets for multiple systems.

9.2.3 Goal of the Validation

We define the following GQM goal for the validation of MTCC:

Goal Validate

Focus the suitability for automatic testing of a Digital Library system family of

Object the MTCC approach

Viewpoint from the perspective of domain experts.

We already discussed the goal of validation in Section 9.1. Since the validation is done from the viewpoint of domain experts, the usability of the MTCC editor and more generally the comprehensibility of the MTCC concepts are of special interest to the validation. The object of validation is the implementation of the MTCC approach for Digital Libraries and Solr prepared for this work. While a validation of MTCC independent from any concrete implementation would be optimal, the resources available for this work only allow one implementation to be evaluated. MTCC is a test automation approach. For the purpose of this work, it is applied to the digital library domain. The focus of the GQM goal is therefore on the applicability of MTCC to the Digital Library domain.

9.2.4 Validation of Testing Approaches

MTCC is a testing approach that depends on domain experts for the identification and construction of tests. This results in differences to other approaches to automatic testing. These differences have to be considered in a validation. As we discussed in Section 1.1.3, a number of different, sometimes conflicting, quality attributes can be applied to software for test automation [Kan03, GF00]. If the effectiveness and the efficiency of a testing approach are considered as its most important attributes [Kan03], then the quality of a test is defined by its ability to find faults with minimal resource usage.

In reality, only considering effectiveness and efficiency for a testing approach is an oversimplification. The individual quality attributes of each system have to be evaluated in the context of its goals and the domain [DER05, EBI06] it is applied to. In the MTCC approach, the effectiveness of testing is determined by the ability of the domain experts to identify and design suitable tests for a system. A type III evaluation of MTCC would have to include domain experts and assess the effectiveness of their tests and the efficiency of the MTCC test automation over a long period of time and under realistic conditions. A validation of such scale is not possible in the scope of this work.

We focus our validation on the feasibility and practicality of the approach, especially on the ability of MTCC to express the necessary tests for the domain model, instantiate an editor instance for these models that allows the construction of tests configurations, and use the Test Configurations for testing the system.

9.3 GQM Application for MTCC

In the following, we define five questions related to the goal defined in the previous Section 9.2. The questions allow the assessment of the degree to which MTCC supports the goal that we derived from our hypothesis. An overview of all questions, subquestions and metrics is given in Table 9.1. The feasibility of the GQM approach is addressed by the following three questions:

Q1 Can all relevant tests for the domain experts be represented by MTCC Test Configurations?

Q2 Is it possible to generate test cases from the abstract MTCC test configuration models and to execute these tests on the systems under test?

Q3 Can Test Configurations that were constructed for one system be reused for another system?

Question Q1 considers the basic feasibility for automated testing — MTCC can only automate a test that can be expressed in terms of MTCC models. Question Q2 builds on question Q1. If MTCC can represent a particular test in a model, it does not necessarily mean that the models can be used for the generation of test cases. Question Q3 also depends on question Q1 — given that a test can be expressed for one system, is it possible to reuse the test — unchanged after a transformation — for another system?

The fourth MTCC question considers the type II validation of MTCC. The question validates the practicality of the MTCC approach. We examine whether domain experts understand the concepts of the MTCC approach as the are presented by the MTCC editor.

Q4 Are the MTCC models represented by the editor suited for test construction by domain experts?

Only a limited type III validation is conducted for MTCC. Question 5 examines the time necessary for the manual execution of tests and compares this with the resources necessary to execute tests in MTCC.

Q5 What is the relationship between the resources necessary for the manual execution of tests compared to the resources needed to automate and execute the same tests with MTCC?

9.3.1 Systems and Test Sets

For the following discussion of the MTCC validation plan, it is necessary to define several sets, variables and functions.

Our evaluation considers the system family S of four systems $S = \{s_1, s_2, s_3, s_4\}$. The systems are discussed in detail in Chapter 8.

s_1 The search of the Sowiport portal.

s_2 The interdisciplinary portal infoconnex

s_3 The portal for international relations and area studies of the German Institute of International and Security Affairs

s_4 The search server Apache Solr

In order to assess the capability of MTCC to represent relevant tests for a domain, we use a number of different test sets. A test set is a collection of tests or testable faults for one of the members of the system family S. Section 9.3.2 discusses test sets in detail.

TS_1 This test set includes a number of tests that we defined as a baseline for the MTCC approach before the approach was implemented.

TS_2 This test set includes a number of test scenarios for quality attributes from the Information Retrieval domain that can be applied to the system s_4.

TS_{3a} A set of tests that a domain experts defined in a simple text format. This test set addresses the system s_1.

TS_{3b} A set of tests that was specified in the same text format as the tests in the test set TS_{3a}. In contrast to TS_{3a}, the test in the test set address system s_3 and were defined by domain experts for this portal.

TS_{4a} This test set includes a number of tests that are defined in a test plan for the manual testing of system s_2.

TS_{4b} This test set includes tests for system s_2. Like the tests in test set TS_{4a}, the tests were originally defined in a test plan for manual execution. Unlike the tests in TS_{4a}, the tests only address the search functionality of s_2.

T_5 This test set does not include tests as such but rather faults and TODO items that were reported for system s_1. We treat this faults as test scenarios if a regression test could be derived from the fault .

Each test set contains a different number of tests. The number of tests for each test set is stored in the variables $t_{x,y}x \in \{1,2,3a,3b,4a,4b,5\}, x \in \mathbb{N}$

Question 1	Is MTCC capable to represent all relevant tests?
Subquestion 1.1	Does MTCC support predefined test scenarios?
Metric 1.1.1	Degree of support for test set TS_1
Subquestion 1.2	Is MTCC capable to express Information Retrieval tests?
Metric 1.2.1	Degree of support for test set TS_2
Subquestion 1.3	Can MTCC express tests defined by domain experts?
Metric 1.3.1	Degree of support for test set TS_{3a}
Metric 1.3.2	Degree of support for test set TS_{3b}
Subquestion 1.4	Can MTCC represent existing manual tests?
Metric 1.4.1	Degree of support for test set TS_{4a}
Metric 1.4.2	Degree of support for test set TS_{4b}
Subquestion 1.5	Does MTCC support tests derived from identified faults?
Metric 1.5.1	Degree of support for test set TS_5
Question 2	Can MTCC Test Configurations be executed as automated tests?
Subquestion 2.1	Is a near-complete implementation of the execution system for one testee possible?
Metric 2.1.1	Percentage of executable Test Steps for TS_1 and TS_{3a}
Metric 2.1.2	Lines of code needed to make TS_1 and TS_{3a} executable
Subquestion 2.2	Can MTCC Test Configurations be executed on all members of a system family?
Metric 2.2.1	Percentage of executable Test Steps for t_{a1} for the system family S
Metric 2.2.2	Lines of code needed to make t_{a1} executable on all systems in S
Subquestion 2.3	Can MTCC Test Configuration be executed with different test runners?
Metric 2.3.1	Percentage of executable Test Steps for test t_{a1} on system s_1 and the test runners tr_1 and tr_2
Metric 2.3.2	Lines of code needed to make t_{a1} executable on system s_1 and the test runners tr_1 and tr_2
Subquestion 2.4	Is MTCC compatible with existing testing regimes?
Question 3	Can MTCC Test Configurations be reused?
Subquestion 3.1	Are Test Configurations reusable for system variants?
Metric 3.1.1	Percentage of the reusable tests for system s_3 supported in MTCC
Subquestion 3.2	Are Test Configurations reusable for systems versions?
Metric 3.2.1	Percentage of the reusable tests for system s_2 supported in MTCC
Subquestion 3.3	Are Test Configurations reusable for system layers?
Metric 3.3.1	Percentage of the reusable tests for system s_4 supported in MTCC
Question 4	Is MTCC practical and understandable by domain experts?
Subquestion 4.1	Can domain experts model and adapt simple tests?
Metric 4.1.1	Time needed by the domain experts to model $t_{4b,88}$ and adapt the test to TV_1
Subquestion 4.2	Can domain experts construct new tests?
Metric 4.2.1	Help needed by the domain experts to model TV_{2a} and TV_{2b}
Question 5	Is MTCC efficient compared to manual testing?

Table 9.1: GQM questions and metrics used in the MTCC validation

9.3.2 Q1: Capability to Represent Tests

Question Q1 of the MTCC validation is: Can all tests that are relevant for the domain experts of the current system family be expressed by MTCC models? In order to answer this question we test to which degree the tests defined in the different test sets introduced in Section 9.3.1 can be represented by the MTCC models for the respective systems.

For every test $t \in T$ with $T \in T_{gesamt}, T_{gesamt} = \{TS_1, TS_2, TS_{3a}, TS_{3b}, TS_{4a}, TS_{4b}, TS_5\}$ the degree to which the test can be represented in MTCC is represented by the value gr, gr is defined as follows.

$$gr \in G, G = \{g_{\text{yes}}, g_{\text{can}}, g_{\text{no}}, g_{\text{extern}}\} \tag{9.1}$$

g_{yes} The test can be completely represented by the MTCC models used for the validation. This means that the Application Feature Model, the Application State Model, and the Domain Test Model are complete for the purpose of the test.

g_{can} The test cannot be represented with the models used for the validation, but the models could be extended to support the test if the necessary Services, Service instances or Test Steps are added to the Domain Feature Model, Application Feature Model, or Application State Model respectively.

g_{no} The test can neither be represented by the models used for the validation nor is the addition of Services, Service instances or Test Steps possible that would allow the representation of the test.

g_{extern} The test is not within the scope of MTCC. Examples for such tests are tests that are specific to the graphical user interface of a testee.

We consider a test t as supported by the MTCC approach if $gr(t) = g_{\text{yes}}$. When a test set $TS \in TS_{gesamt}$ is considered in its entirety, then the support of MTCC for this test set is defined as

$$gr(T) = \frac{|\{t_1 : t_1 \in T \wedge g(t_1) = g_{\text{yes}}\}| * 100}{|\{t_2 : t_2 \in T \wedge g(t_2) \in \{g_{\text{yes}}, g_{\text{can}}\ g_{\text{no}}\}|} \tag{9.2}$$

9.3.2.1 Q1.1 Support of MTCC for Predefined Test Scenarios

The test set TS_1 has been defined prior to the design and implementation of the MTCC approach. The test set consists of 16 tests for system s_1. The test set was defined to serve as the basic for the development and validation of the MTCC models and was the result of a first informal analysis of the testable functionality of the testee. Each of the 16 tests covers a different functionality of the system and can therefore be considered as a representative of a test class. We define a test class as a trace through the functionality of the system for which many instances can exist. Searching for a term and comparing the number of result documents with an expectation would be a test class for which search for concrete terms would be instances or tests.

Each test in TS_1 is described as a test scenario in a structured test format that consists of actions taken on the interface of the testee and the expectation or assertions that must be true. An example of such a test description is given below. We translated it from the original German:

```
Test 4(bs) Test of the SOLIS database filter

This tests verifies the filters for databases for the extended
 search in Solis

  1. Action The filter for databases is selected
      in the Advanced Search
  2. Action The Check Box all is de-selected
  3. Action The Check Box Solis is selected
  4. Action The term 'social' is entered in
     the text field of the search form
  5. Assertion More than 17.000 documents are found.
  6. Assertion The person navigator contains
     30 hits for the value Weber, Max
```

Every test scenario has a name and a short description. The test scenario is represented as a sequence of interactions that take place between the user and the system under test. Since user and system interact via the user interface, the actions refer to elements of the GUI. Actions and assertions in the test scenario correspond to the two action types and assertions in MTCC.

We choose the above format for the representation of test cases since it is precise enough to serve as a basis for the manual execution of tests and test modeling while it is not bound to a particular modeling language.

We argue that test set TS_1 is relevant for the validation of MTCC since it covers a wide range of functionality for the system s_1 and is independent of the MTCC approach. Metric M 1.1.1 describes $g(TS_1)$, the degree of support for the tests of the test set TS_1.

9.3.2.2 Q1.2 Capability to Express Information Retrieval Tests

The test set TS_2 consists of tests that assess the results of applying cross-concordances [MP08] to the retrieval process. The test set holds the 7 tests below that verify various expectations about the use of cross-concordances.

$t_{2,1}$ Are more documents found for a query that was expanded with cross-concordances than for queries that were not expanded?

$t_{2,2}$ Do the 10 or 20 most relevant documents for a query change when cross-concordances are used?

$t_{2,3}$ Does the ranking of the most relevant 10 or 20 documents change?

$t_{2,4}$ Do cross-concordances result in documents form more databases to be included in the result set than would be the case without cross-concordances?

$t_{2,5}$ How do cross-concordances change the number of databases in the most relevant 10, 20 or 100 documents?

$t_{2,6}$ How does the recall change for 10, 20, and 100 documents when cross-concordances are used?

$t_{2,7}$ How do cross-concordances influence the precision for the first 10, 20, and 100 documents of a result list?

Each of the 7 tests above would be tested with a initial query, for example a query derived from a CLEF [PBS07] topic.

Question Q1.2 assesses of the ability of MTCC to express tests for the effectiveness of Information Retrieval systems. Since the ability of a system to return documents that are relevant to an information need cannot be completely automated, MTCC must support tests that calculate recall and precision passed on pooled relevance assessments and must be able to test hypotheses like those expressed in the seven tests above.

Metric M1.2.1 describes the degree to which the test set $g(TS_2)$ can be expressed with MTCC

9.3.2.3 Q1.3 Capability to Express Tests Defined by Domain Experts

In the preparation for this work, domain experts affiliated with the Digital Libraries s_1 and s_3 were asked to specify tests for their respective system in the format described in Section 9.3.2.1. The domain experts were informed that the tests were collected in order to use them in a test automation approach, but did not know about MTCC. Examples for the specified tests can be found in Section A and Section A in the appendix.

Compared to the tests in test set TS_1, the tests in the test sets TS_{3a} and TS_{3b} were not defined for the evaluation of the MTCC approach but represent actual requirements for either system s_1 or s_3. Since the tests in the test sets are defined by domain experts, we argue that they are likely to represent realistic requirements to a test automation system.

The metrics M1.3.1 and M1.3.2 describe $g(TS_{3a})$ and $TS_{3b})$, the degree to which MTCC supports the expression of tests for the respective test sets.

9.3.2.4 Q1.4 Representation of Existing Manual Tests

The test sets TS_{4a} and TS_{4b} consist of manual tests for system s_2 that execute monthly on the system. Section A in the appendix gives an example of the tests in the test sets. A particularity of the tests in these test sets is that they do not define the expected behavior of the system under test, only the actions to be taken on the system.

Test set TS_{4a} consists of tests for different aspects of system s_2, some of which address details of the GUI and are therefore not in the scope of MTCC. All tests in the test set TS_{4b} address the search functionality of the system. Compared to the tests described in TS_1, TS_{4b} does not cover much of the functionality of the system under test but instead tests the search functionality of system s_1 in detail. The fact that all tests in TS_{4b} are instances of only a few test classes is relevant for the validation since it is likely that MTCC either supports all tests of a test class or no tests from this test class.

We argue that the test set TS_{4b} is relevant for the validation because it represents real tests taken from practice.

Metric M1.4.2 describes $g(TS_{4b})$, the support of MTCC for the tests in test set TS_{4b}, metric M1.4.1 considers $g(TS_{4a})$.

9.3.2.5 Q1.5 Support for Tests Derived from Identified Faults

Test set TS_5 does not include tests as such. Rather, the test set is an export of all entries of the issue tracking system for system s_2. Some of the entries in the issue tracker are feature requests and many of the remaining faults in the issue tracker address aspects of the system that are not within the scope of MTCC. These entries are assigned to the category g_{extern}. The subset of entries in the issue tracker that are in the scope of MTCC are faults that where found by domain experts in ad-hoc testing. Since it is desirable to automate such ad-hoc tests with MTCC, we derive tests from the identified faults and assess if the tests can be expressed with MTCC.

Metric M1.5.1 describes $g(TS_5)$, the degree to which MTCC supports these tests.

9.3.3 Q2: Executability of MTCC Test Configurations

The second question of the type I validation examines the executability of the MTCC Test Configuration models. We consider a Test Configuration instance executable when it can be used — either in its original form or after a transformation — to exercise a testee. Since Test Configuration instances serve as the basis for test case generation, a model is executable if a test case can be created from the model.

The question Q2 has a quantitative aspect. It is not only of interest if test code can be generated from the models but also how much resources are necessary to establish a test code generation infrastructure and to use this infrastructure to generate test cases.

We call the infrastructure needed for the generation of test cases and for the execution of these test cases the execution system. The execution system consists of the Configuration Reader, the Test Generator and the Test Adapter.

With the consideration of the resources necessary for test code generation and execution, question Q2 seems to address the efficiency of MTCC and thereby the type III validation. We argue that Q2 is part of the type I validation because the feasibility of MTCC is only given if MTCC can be implemented with the resources available for test automation in a given project.

Question Q2 consists of 4 subquestions:

- To what degree is the complete implementation of the execution system possible for a given testee? In order to answer this question, we implement an execution system for the tests of test sets T_1 and T_{3a} on system s_1. The question examines how many resources are necessary for the implementation of an execution system that covers most of the functionality of a system.

- Is the implementation of an execution system feasible for all members of a system family? In order to examine this question we implement a limited execution system for all members of the system family. This execution system only supports the execution of the single test class t_{a1}.

- Is the implementation of an execution system for different test runners possible? We implement an execution system for the test class t_{a1} on system s_1

with the test runner tr_2. This question investigates whether the MTCC test models are specific to one particular test runner.

- Can the test cases generated by the MTCC execution system be integrated into an existing testing process?

In order to investigate the above questions, we implement the needed execution systems for the MTCC prototype. We document the lines of code for each implementation as a simple metric for the resources necessary to implement the system. As discussed in Section 10.7.1, we consider lines of code an imperfect but usable metric.

9.3.3.1 Q2.1 Near-Complete Implementation for a Testee

This question examines the resources necessary for a nearly complete implementation of the execution system of MTCC for one system. We call the implementation near-complete since it implements all tests that are defined for a system, not all functionality of the system. We implement the test sets T_1 and T_{3a} for system s_1.

Metric M2.1.1 describes the percentage of all tests from T_1 and T_{3a} that can be supported by the implemented execution system. Metric M2.1.1 investigates the lines of code that are used for the implementation of the parts of the execution system.

9.3.3.2 Q2.2 Executability on the Members of a System Family

This subquestion addresses the executability of MTCC Test Configurations on the different members of the considered system family. These questions aims to prove that MTCC Test Configurations can be executed on all members of a given system family. In order to answer the question, an execution system for the test t_{a1} is implemented for all system in S. t_{a1} includes the following steps:

```
1. Action A search using multiple fields in an advanced
     search form is started.
2. Action The number of result documents is stored in
   a variable.
3. Assertion The number in the variable is compared
   to some expected value.
```

Test t_{a1} has been chosen for the validation since all systems in the system family S support the functionality required by the test. We restrict the validation of this question to a single test because of the available resource for this validation and since the test class described by test t_{a1} is the most frequently used for this system family as can be seen from T_{4a}.

Metric M2.2.1 examines whether the execution system for t_{a1} can be implemented for all members of the system family. Metric M2.2.2 describes the lines of code necessary for these implementations.

9.3.3.3 Q2.3: Executability with Different Test Runners

Question Q2.3 addresses the executability of MTCC Test Configurations with different test runners. Two test runners are used for the validation.

The test framework Selenium (tr_1) is used for testing a system by automating a web browser. Twill (tr_2) is a HTTP Protocol Driver [BGH07] used for testing web applications.

A test execution system is implemented for test t_{a1} on the system s_1 and for both test runners. Metric M2.3.1 describes whether the implementation of the execution system succeeds for the two test runners. Metric M2.3.2 measures the lines of code necessary for the implementation.

9.3.3.4 Q2.4: Compatibility with Existing Testing Regimes

Question Q2.4 addresses the compatibility of the test cases generated for the questions Q2.1, Q2.2, and Q2.3 with existing testing tools and the degree to which MTCC test cases can be integrated in existing testing regimes. For this question we examine the infrastructure needed in order to execute MTCC tests. This is a qualitative question that is not based on any metrics.

9.3.4 Q3: Reuse of Test Configurations

Question Q3, the third question of the type I validation, examines to what extent Test Configuration instances can be reused for different systems. Specifically, we investigate whether a Test Configuration instance that is specific to one Application Test Model can be transfered to the Application Test Model of another system. If such a transfer of test models is possible and if execution systems exist for all testees, then tests are reusable for different systems.

Question Q3 consists of three sub questions each of which corresponds to one member of the system family S for which tests can be reused. For each of the systems and each test in the test sets T_1 and T_3a that is intellectually reusable we check whether a Test Configuration created for system $s1$ can be reused for the target system. We call a test reusable by hand if a domain expert with knowledge of the original test can execute the scenario described by the test on another system, that is, if all features required by the test are present for both systems and the requirements expressed by the test can therefore be verified on both systems.

9.3.4.1 Q3.1: Reuse for System Variants

Question Q3.1 considers the reuse of Test Configuration instances for variants of a testee. Systems s_1 and s_3 were developed in parallel and are similar regarding their implementation, we therefore call these systems variants of each other.

For every test $t, t \in T_1$ the function $x = trans(s_1, s_3, t), x \in \{g_{yes}, g_{no}, g_{can}, g_{false}\}$ describes whether a Test Configuration can be transfered or not.

The value g_{no} is used for tests that are not reusable by hand for different systems. Tests with the value g_{can} are reusable by hand but cannot be reused by MTCC. The value g_{yes} is used for tests that are reusable and whose semantics are correctly represented for the systems for which the test is reused. The value g_{false} is used for tests that can be automatically transfered but whose semantics are changed by the transfer process — the test for the target system is not a correct representation of the test on the original system.

Metric M3.1.1 describes the percentage of the reusable tests that can be reused in MTCC.

$$M3.1.1 = \frac{|\{t_1 : t_1 \in T_1 \wedge trans(s_1, s_3, t_1) = g_{yes}\}| * 100}{|\{t_2 : t_2 \in T_1 \wedge trans(s_1, s_3, t_2) \in \{g_{yes}, g_{can}\ g_{false}\}|}. \tag{9.3}$$

9.3.4.2 Q3.2: Reuse for Systems Versions

Question Q3.2 considers the reuse of test models for a system and another system that is the precursor of the first system. System s_2 can be considered a precursor of system s_1 since s_2 was developed earlier and both systems offer the same basic functionality, even though system s_2 includes some functionality that is not part of s_1.

Metric M3.2.1 for question Q3.2 is analogous to metric M3.1.1 but considers the transfer of tests between the systems s_1 and s_2.

9.3.4.3 Q3.3: Reuse for System Layers

Question Q3.3 examines the reuse of Test Configurations that were constructed for a system and are transfered to a subsystem. Specifically, we test if tests for system s_1 can be reused for system s_4. Since system s_4 is a search server not a

Digital Library, these tests assess the potential for reuse between different layers of a system.

Metric M3.3.1 is analogous to metrics M3.2.1 and M3.1.1, but considers system s_4.

9.3.5 Q4: Practicality and Understandability of MTCC

Question Q4 considers the practicality of MTCC. More precisely, it investigates whether domain experts understand the approach and can use the editor to build Test Configurations from given test scenarios. Question Q4 is part of the type II validation.

The question primarily investigates whether the MTCC is understandable by domain experts and can be applied by them. While the usability of the editor implementation used for the validation contributes to the practicality, it is not the focus of the validation.

We investigate the practicality of MTCC based on two tasks, the adaptation of existing Test Configurations and the construction of new Test Configurations from textual specification. We consider quantitative as well as qualitative aspects. Qualitative aspects include feedback given by the domain experts about the usability of the editor and the understandability of the MTCC approach as well as the most frequent faults and problems that occur in using the editor. The quantitative aspects measure if the domain experts were able to express the scenarios without help, with some help or with no help at all.

We use test sets with seven tests each for the validation of the practicality of the approach. One type serves as the basis of the construction of new tests, the other consists of tests used to validate the adaptation of existing tests.

The test sets TV_{2a} and TV_{2b} each contain seven tests that were randomly selected from the test sets T_1 and T_{3a}. The test sets hold test scenarios that are used to validate the construction of new tests by domain experts.

$$TV_{2a} = \{ts_{1,10}, ts_{1,11}, ts_{3a,0}, ts_{3a,10}, ts_{3a,15}, ts_{3a,28}, ts_{3a,5}\} \tag{9.4}$$

$$TV_{2b} = \{ts_{3a,11}, ts_{3a,12}, ts_{3a,14}, ts_{3a,17}, ts_{3a,21}, ts_{3a,23}, ts_{3a,8}\} \tag{9.5}$$

The test set TV_1 consists of five tests randomly selected from the test set TS_{4b}. The tests in the test set are all based on the same test class: The submission of a query and the calculation of the results of a query.

$$TV_1 = \{t_{4b,88}, t_{4b,89}, t_{4b,90}, t_{4b,103}, t_{4b,104}\} \tag{9.6}$$

Four domain experts were involved in the validation, we denote them by the variable names $de_{1...4}$.

9.3.5.1 Q4.1: Adaptation of Tests

Question Q4.1 investigates the adaptation of Test Configurations from test set TS_{4B} by domain experts. Both the time needed for the adaptation of the tests and potential problems during the adaptation are considered. We argue that the adaptation of existing tests and the creation of new instances for already existing test classes is a frequent activity in testing and that tests from the test set TS_{4B} are well suited to represent this activity.

Metric M4.1.1 measures the average time needed by the domain experts to construct the test $t_{4b,88}$ and to adapt this test to the scenarios described in the remaining entries of the test set.

9.3.5.2 Q4.2: Construction of Tests

Question Q4.2 investigates the degree to which domain experts can use the MTCC editor to construct tests from given test scenarios. The question considers 14 tests, subdivided into two subsets of 7 tests each. In order to validate the practicality of MTCC, four domain experts evaluated the MTCC editor in 2 sessions each. During the first session, the tests from test set TS_{2a} were investigated. In the second session, the domain experts were given the task to model the tests from TS_{2b}. Details are given in Section 10.6.1.

As a quantitative measure we determine the degree to which the domain experts can express the test with the MTCC editor. We differentiate three degrees of support that can be given to a domain experts during test construction. If a test is modeled without any help the test falls into the category $nohelp$.

If some help is needed, either with respect to the MTCC approach or the MTCC editor, the respective tests falls into the category $help$. Examples for the help given to the domain expert are answers to concrete questions about the user interface or the meaning of the parameters for specific Test Steps. Generally, all information that could also be learned from the manual is considered limited help. If more help is needed and if it is likely that the construction of the test would fail even with the use of the manual, the test is assigned to the category $fullhelp$.

Details about the categories can be found in Section 10.6.1 Metric M4.2.1 describes the distribution of the tests TV_{2a} and TV_{2b} on the categories $nohelp$, $help$, and $fullhelp$ for all four participating domain experts.

9.3.6 Q5 Efficiency of MTCC

Question Q5 considers the efficiency of MTCC in comparison to manual testing. The question is an entry point to a further type III validation at a later date. We consider manual testing since it is the current approach used for testing in validating organization. A comparison of MTCC to other approaches of automated testing is the subject of later work.

The purpose of question Q5 is to assess the resources necessary for manual testing and testing with MTCC. As discussed in Section 1.1.3, four phases of testing can be differentiated. For question Q5 we only consider the third phase of testing, the execution of tests.

We compare the time needed for the manual execution of the tests in the test set TS_{4b} on system s_2 with the resources necessary to implement the execution system for the tests and the system. We argue that the comparison of manual testing with implementation costs for the test generation and execution system provides a better measure than the comparison with the resources necessary for test case generation and test execution since the automatic test execution does not depend on human involvement but does depend on the potentially resource-intensive implementation of the execution system.

We use two measures to discuss the efficiency of MTCC, (1) the lines of program code necessary to implement the execution system and (2) the time needed for the manual execution of the tests in test set TS_{4b}. The manual execution of tests is done biweekly.

9.4 Chapter Summary

Software Engineering is an empirical science, MTCC must therefore be evaluated in a systematic manner. This work applies the Goal Question Metric (GQM) approach to validate the feasibility and practicality of MTCC. The MTCC validation is conducted as a case study. The validation of MTCC consists of a type I and type II validation. The type I validation addresses the basic feasibility of the approach, the type II validation investigates the practicability the the approach. The type III validation compares the efficiency of MTCC with the efficiency of other approaches.

The type I validation of MTCC uses three questions: The first question assesses the capability of MTCC to represent tests that are relevant to domain experts. The second question investigates how abstract Test Configurations can be used the exercise concrete systems. The third question of the type I validation investigates the effectiveness of the methods employed by MTCC to reuse tests. The type II validation consists of one question that investigates the understandability of the MTCC approach to domain experts. The type III validation investigates the necessary resources for testing with MTCC with the resources necessary for manual testing.

10 Results of the Validation

The validation of MTCC was conducted between March and May 2008 at the GESIS [1] in Bonn. The participants of the validation were employees of the GESIS who were either involved in the specification of the systems, in the design of tests for one of the systems or in the manual testing of the systems. The validation was conducted on systems $s_{1...4}$. With the exception of system s_4, all systems were original systems used in production. The usability of MTCC was validated with a prototype of the MTCC editor.

10.1 Participants in the MTCC Validation

Four domain experts were participating in the MTCC evaluation:

p_1 This domain expert had experience in test design for system s_1 and took part in the manual testing of this system.

p_2 This domain expert was a student employee who undertook the execution of the manual tests defined in the test sets T_{4a} and T_{4b} for system s_2 in the years 2006 and 2007.

p_3 Domain expert p_3 was the student employee who succeeded domain expert p_2 in the execution of the tests for system s_2 in 2008.

p_4 This domain expert defined the test set T_2 and was familiar with the different systems within the system family.

10.2 The Prototype of the MTCC Editor

The implementation of MTCC was done according to the principles and techniques discussed in this work. A class library for the construction and manipulation of feature models implemented for this work was used to instantiate the Domain Test Model and the Application Feature Model and the Application State Model for every SUT.

---

[1] German Social Science Infrastructure Services

Figure 10.1 illustrates the overall architecture of the MTCC prototype. We distinguish five logical packages of functionality: The Domain & System package includes class libraries that support the specification of domain-level and system-level models. These class libraries build on a Feature Modeling Layer that provides an implementation of feature models with cardinalities. The feature models in turn are implemented as nodes in a Document Object Model. The User Interface package includes the Editor that is responsible for the overall lock and feel of the GUI used for test modeling and the GuiBuilder classes that represent individual Test Step instances. The Model Composer is responsible for the instantiation of the Application Test Model and the Reuse System that rewrites Test Configurations for different system models are also part of the Editor package.

The Generator package includes the Code Generator that transforms MTCC Test Configurations into executable test cases. Templates determine the overall structure of the test code. The Test Parameter Writer rewrites the Test Configurations into a simpler representation better suited for code generation.

The MTCC Runtime and the External Runtime packages are responsible for the execution of the generated Test Cases. The Test Adapter provides a platform for test execution the decouples from the implementation details of the Test Runner and the SUT. The Test Runner executes the Test Cases. The Test Runner is often supported by Support Libraries for test execution, this work uses Selenium or twill. The Language Runtime used for test execution can be different from the language in which MTCC is implemented.

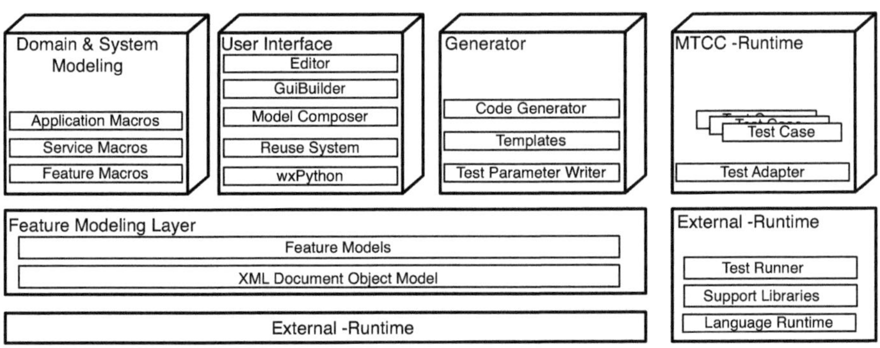

Figure 10.1: Logical architecture of the MTCC prototype

The prototype implementation of MTCC used for the validation was done in Python[2]. XML was used as a serialization format for the feature models and for the state machines used in the validation. wxPython[3] was used to implement the GUI. The implementation comprises 6600 Lines of Python.

Figure 10.2 displays a screenshot of the implementation of the editor as it was used for the validation.

As described in Section 7.1, the left part of the GUI displays the Test Steps that the current test consists of; the right part of the editor represents the currently selected Test Step and allows its parameterization. The validation prototype was tested on Mac OS X and MS Windows systems.

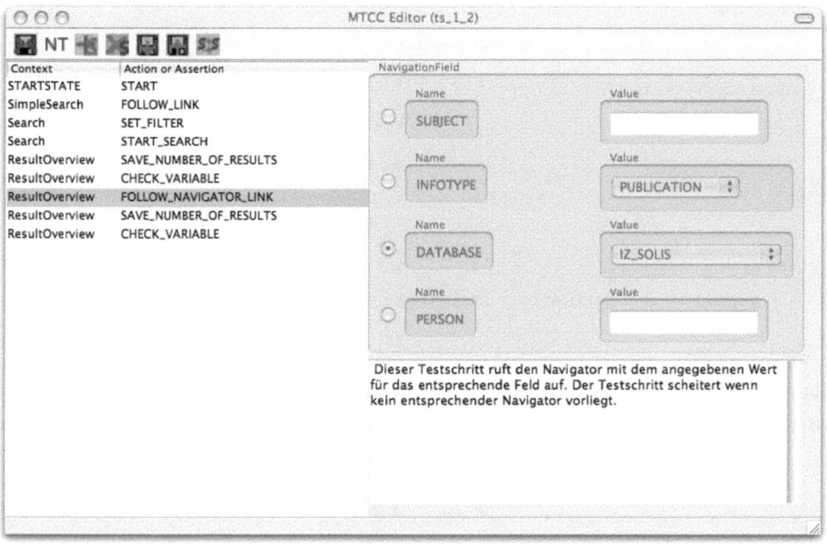

Figure 10.2: Screen shot of the editor prototype used for the validation

The dialog used for the selection of Test Step instances is displayed in Figure 10.3.

The prototype implements the techniques for test reuse discussed in Section 7.3. Figure 10.4 displayed a screen shot of the dialog used by MTCC to allow the selection of Test Step instances when a test is transfered between different systems.

[2]http://python.org

[3]http://wxpython.org

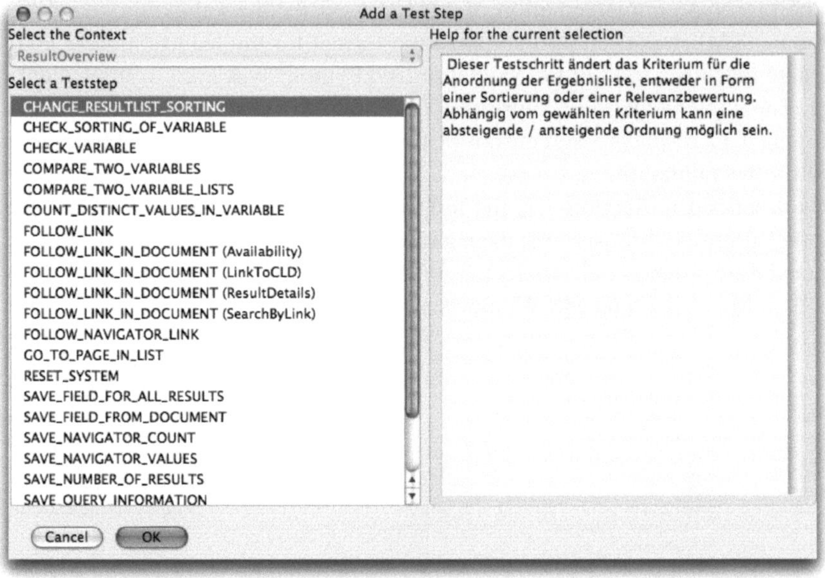

Figure 10.3: Selection of a Test Step instance in the MTCC prototype

Since only one sequence of Test Step instances is available for the tests and systems used in Figure 10.4, the dialog is deactivated.

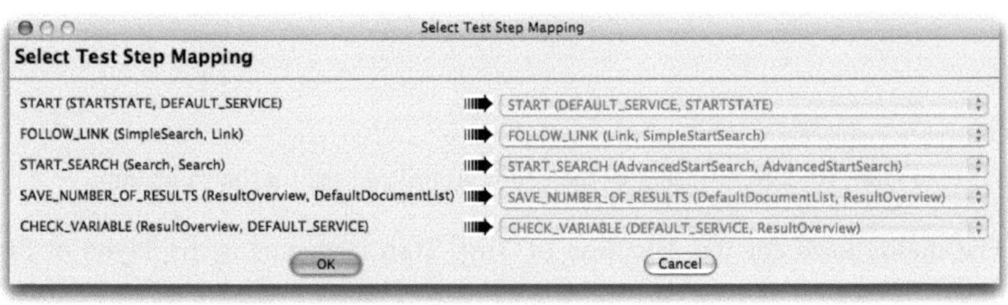

Figure 10.4: Transfer of a Test Configuration to another system in the MTCC editor

10.3 Considered Systems

The validation considered the systems introduced in Section 9.3.1 on page 160. Every system was represented with a specific Application State Model and Application Feature Model; a common Domain Test Model was used for all models. All models were modeled to the extent which was needed for the validation — features that are not exercised by any test set were not included in the models. No changes for the purpose of the validation were made to the systems.

10.4 Capability to Represent Tests

In order to validate MTCC's capability to represent tests for the application domain of Digital Libraries, we assessed for each entry t of every test set with $t \in T, T \in \{T_1, T_2, T_{3a}, T_{3b}, T_{4a}, T_{4b}, T_5\}$ into which category of support $g \in \{g_{\text{yes}}, g_{\text{no}}, g_{\text{can}}, g_{\text{extern}}\}$ the test falls. The total values of the different test sets and categories are displayed in Table 10.1. Table 10.3 displays the values for the metrics of question Q1 as defined in Section 9.3.2.

Test Set	System	Number of Tests	g_{yes}	g_{no}	g_{can}	g_{extern}
TS_1	s_1	16	16	0	0	0
TS_2	s_4	7	7	0	0	0
TS_{3a}	s_1	30	26	3	0	1
TS_{3b}	s_3	50	41	2	3	4
TS_{4a}	s_2	12	8	0	0	4
TS_{4b}	s_2	137	137	0	0	0
TS_5	s_1	352	206	13	3	130
$TS_{1...5}$		604	441	18	6	139

Table 10.1: Support of MTCC for the validation of test sets

The different test sets displayed in Table 10.1 differ considerable both in the number of entries and in the properties of the included tests. Before we discuss how the validation of MTCC was conducted and examine the results of the validation regarding the capability of MTCC to express tests for the application domain, we examine of the test sets $TS_{1...5}$.

We differentiate three groups of test sets: (1) Test sets that were defined for the purpose of test automation, (2) test sets containing tests from current testing practice, and (3) test sets with entries that were derived from the entries of issue tracking systems.

- The test sets TS_1 and TS_2 were defined as basic requirements for a system for the automated execution of acceptance tests. The test sets TS_{3a} and TS_{3b} were defined as manual tests for systems s_1 and s_3. The test sets differ in the functionality of their respective testees and in the specificity of the tests.
- The test sets TS_{4a} and TS_{4b} hold tests that are executed biweekly on system s_2. Test set TS_{4a} consists of a number of different tests for a Digital Library, including a number of tests that are specific to the user interface of the system. The entries of test set TS_{4b} represent different queries for system s_2.
- Test set TS_5 differs from the other test sets in terms of the number of included tests as well as in the origin of the tests. With 352 entries, this test set includes more potential tests than all other test sets taken together. TS_5 is different from the other test sets as it contains faults and outstanding issues taken from the issue tracker of system s_1. Since the entries of the test set were not originally developed as tests, a large number of them are not in the scope of MTCC and fall in the category $\mathrm{g}_{\mathrm{extern}}$.

In the context of this validation we decide for every potential test in every test set if the entry is a test for the functionality covered by MTCC. If this is not the case, we assign the entry to the category $\mathrm{g}_{\mathrm{extern}}$. As discussed in Section 9.3.2 on page 162, we differentiate three categories of support for tests that are within the domain covered by MTCC.

We discuss and interpret the distribution of the entries in the different test sets for the categories of support in Section 10.7.2. In the following, we give an overview of the results of the validation and point out relevant aspects.

Table 10.2 shows the distribution of the tests in the test set over the different categories of support.

MTCC supports 94,84% of all tests if all elements of of the test sets $TS_{1...5}$ that fall into the category $\mathrm{g}_{\mathrm{extern}}$ are excluded from consideration. If the tests in $\mathrm{g}_{\mathrm{extern}}$ are considered, 73.01% of all tests can be represented by MTCC.

All tests of the test sets TS_1, TS_2, and TS_{4b} can be represented by MTCC. The test sets TS_{3a} and TS_{3b} include tests that fall in the category g_{no}.

Of the three tests of test set TS_{3a} that cannot be represented with MTCC, test $t_{3a,16}$ is not supported because the representation of the test would require support for explicit control flow in Test Steps.

Test Set	g_{yes}	g_{no}	g_{can}	g_{extern}
TS_1	100.00%	0.00%	0.00%	0.00%
TS_2	100.00%	0.00%	0.00%	0.00%
TS_{3a}	86.67%	10.00%	0.00%	3.33%
TS_{3b}	82.00%	4.00%	6.00%	8.00%
TS_{4a}	66.67%	0.00%	0.00%	33.33%
TS_{4b}	100.00%	0.00%	0.00%	0.00%
TS_5	58.52%	3.69%	0.85%	36.93%
$TS_{1...5}$	73.01%	2.98%	0.99%	23.01%
$TS_{1...5}$ without g_{extern}	94.84%	3.87%	1.29%	

Table 10.2: Distribution of the entries of the validation test sets over the categories of support

The tests $t_{3a,18}$ and $t_{3a,20}$ as well as $t_{3b,31}$ and $t_{3b,33}$ of test set TS_{3b} can be expressed with MTCC test model. Still, the tests fall in the category g_{no} because the execution of the tests would require access to external systems like the Refworks reference management system or email accounts, neither of which the MTCC execution system supports.

Test set TS_{3a} includes three tests of the category g_{can}. These tests address functionality of the system s_3 that is not represented in the Domain Feature Model. Specifically, the tests address an Service that allows the selection and use of terms from a controlled vocabulary for a search.

The test sets TS_{4a} and TS_5 each contain more than 30% tests that fall in the category g_{extern}. For test set TS_{4a}, these tests primarily address aspects of the specific user interface of system s_2 that are abstracted in MTCC. The entries from test set TS_5 in category g_{extern} are mostly not tests but feature requests.

Table 10.3 displays the metrics for all subquestions of question Q1. For four of the test sets, all examined tests can be expressed with MTCC.

10.5 Validation of the Executability

Question Q2 and its subquestions examine the executability of MTCC test configurations. Question Q2.1 investigates the resources necessary to implement an execution system for system s_1. Questions Q2.2 and Q2.3 examine a partial implementation of the execution system in order to prove the executability of Test

Metric	Test Set T	$gr(T)$
M1.1.1	TS_1	100,0%
M1.2.1	TS_2	100,0%
M1.3.1	TS_{3a}	89,66%
M1.3.2	TS_{3b}	89,13%
M1.4.1	TS_{4a}	100,0%
M1.4.2	TS_{4b}	100,0%
M1.5.1	TS_5	92,79%

Table 10.3: Metrics for question Q1 — percentage of the tests that can be represented in MTCC

Configurations for different testees and test runners. Question Q2.4 investigates to what degree the execution of MTCC test configurations is decoupled from the specifics of the MTCC approach and can be integrated into an existing testing regime.

The tables 10.4 and 10.5 display the metrics for question Q2. The execution of the Test Configurations was possible for all systems and all test runners. The number of lines of code considers only the parts of the execution system that were specifically implemented for the respective validation questions, common infrastructure code is ignored.

Metric	Percentage of executable Test Steps
M2.1.1	100%
M2.2.1	100%
M2.3.1	100%
M2.4.1	100%

Table 10.4: Metrics for question Q2 - percentage of executable Test Steps

All questions consider quantitative as well as qualitative aspects, the qualitative results can be found in Section 9.3.3.

Details about the design and implementation of the MTCC execution system are given in Section 7.4, interesting aspects of individual implementations are discussed in the following.

Question Q2.1 examines if the complete implementation of an execution system for the tests TS_1 and TS_{3a} for system s_1 is feasible. We argue that the implemen-

Metric	Lines of codes
M2.1.2	530
M2.2.2	189
M2.3.2	66

Table 10.5: Metrics for question Q2 - use of resources

tation of the test sets is sufficient to prove the feasibility of a complete implementation of the MTCC execution system since TS_1 and TS_{3a} cover all relevant test classes for system s_1 as defined by the domain experts.

The implementation of all Test Steps of the test sets TS_1 and TS_{3a} for system s_1 and the Selenium test runner was successful. Metric M2.1.1 therefore has a value of 100%, as is displayed in Table 10.4.

Metric M2.1.2 measures the lines of code and thereby the absolute resources necessary for the implementation of the execution system for TS_1 and TS_{3a}. Table 10.5 displays the results.

Table 10.6 displays the lines of code for the complete implementation of the MTCC execution system for all systems and test sets. The function ti returns the lines of code necessary for system s, the test runner tr and the test set ts. Test t_{a1} was defined for question Q2 in Section 9.3.3, it represents a search with a subsequent comparison of the actual and expected number of results.

Execution System	Number of Test Steps	Lines of Code
Configuration Reader	0	261
Test Generator	0	215
Types and Exceptions	0	120
Support Code	0	596
System Independent Test Steps	9	140
$ti(s_1, tr_1, TS_1 \cup TS_{3a})$	23	530
$ti(s_2, tr_1, \{t_{a1}\})$	3	69
$ti(s_3, tr_1, \{t_{a1}\})$	3	55
$ti(s_4, tr_3, \{t_{a1}\})$	5	65
$ti(s_1, tr_2, \{t_{a1}\})$	5	66
Sum	48	925

Table 10.6: Lines of code for the parts of the execution system

The execution system for question Q2.1 was implemented in 670 lines of Python code. 140 lines of this code implement system independent Test Steps that are used for all systems of the system family. 530 lines of code are specific to system s_1 and Selenium. 670 lines of code were necessary to implement the execution system, this is also the value for Metric Metric M2.1.3 measures 20 lines of code that are needed on average to implement each of the 32 Test Step instances needed for the tests defined in TS_1 and TS_{3a}. The implementation of the MTCC execution system for the test tv_1 for different testees and different test runners as addressed in questions Q2.2 and Q2.3 was successful. Table 10.6 displays the results for metrics M2.2.1 and M2.3.1.

```
import unittest
import s1_executor_selenium as runner
from runner_shared import Bunch, Context

class PortalTest(unittest.TestCase):
  def test(self):
    context=Context()

    para = Bunch({})
    self.failUnless(runner.start(context, para))

    para = Bunch({'next_context': 'Search',
          'current_context': 'SimpleSearch'})
    self.failUnless(runner.follow_link(context, para))

    para = Bunch({'searchfields':
          [Bunch({'term_relation': 'AND', 'field_relation': 'AND',
              'name': 'TXT', 'value': '20060111223'})]})
    self.failUnless(runner.start_search(context, para))

if __name__ == "__main__":
  unittest.main()
```

Listing 10.1: Excerpt from a generated test case

The generation of test cases from Test Configuration instances is done as described in Section 7.4. A test case is an instance of a xUnit test class, in Listing 10.1 a Python unittest class.

The generated test cases are dependent on the execution system for the system under test and the test runner, but are independent of the structure of the Test Configuration models used for the generation of test code.

In lines two and three of the listing, support classes and functions from the Test Adapter library are imported. With the exception of the Test Adapter for the tested system and test runner and some general support code, generated test cases are independent from the MTCC approach. The test cases can be executed like tests that were not generated by the MTCC approach and can therefore be integrated in an existing testing approach.

10.6 Validation of Reusability

The reusability of Test Configurations from one system to another was examined by modeling the tests of the test set TS_1 for the system s_1 and testing whether the resulting Test Configurations are transferable to systems s_2, s_3, and s_4.

The tests were not modified or selected with reuse for different systems in mind, a significant number of tests are therefore not reusable regardless of the test transfer methods implemented in MTCC. 50% of the tests in the test set can be reused for system s_4 by applying the techniques described in Section 7.3. 56.25% of the tests are transferable for systems s_2 and s_3. In the case of s_2, three of the transfered tests do not represent the intent of the original test.

Test	s_4	s_3	s_2
$t_{1,0}$	g_{yes}	g_{can}	g_{false}
$t_{1,1}$	g_{no}	g_{yes}	g_{false}
$t_{1,2}$	g_{can}	g_{can}	g_{can}
$t_{1,3}$	g_{yes}	g_{no}	g_{no}
$t_{1,4}$	g_{yes}	g_{yes}	g_{false}
$t_{1,5}$	g_{yes}	g_{no}	g_{no}
$t_{1,6}$	g_{no}	g_{yes}	g_{yes}
$t_{1,7}$	g_{yes}	g_{yes}	g_{no}
$t_{1,8}$	g_{no}	g_{yes}	g_{no}
$t_{1,9}$	g_{no}	g_{no}	g_{yes}
$t_{1,10}$	g_{yes}	g_{yes}	g_{yes}
$t_{1,11}$	g_{no}	g_{yes}	g_{yes}
$t_{1,12}$	g_{yes}	g_{no}	g_{no}
$t_{1,13}$	g_{yes}	g_{yes}	g_{yes}
$t_{1,14}$	g_{no}	g_{yes}	g_{yes}
$t_{1,15}$	g_{no}	g_{no}	g_{no}
Transferability	50%	56,25%	37,5%

Table 10.7: Reusability of tests

Table 10.7 presents the transferability for every test of test set TS_1. As in the discussion of executability of tests, we distinguish different categories of support for the transferability of a test: g_{yes}, g_{no}, and g_{can}. A pair (t,s) with $t \in TS_1, s \in \{s_2, s_3, s_4\}$ falls into the category g_{yes} if a transfer of the test t from system s_1 to system s is possible and is automatically supported by MTCC. If a transfer of the test is possible in theory but is not supported by MTCC for technical reasons, the pair falls in the category g_{can}. The category g_{no} is used when a test t cannot be transfered, neither by automatic or by manual means.

The tests $t_{1,0}$, $t_{1,1}$ and $t_{1,4}$ for system s_2 fall under the category g_{false}. The Test Configurations of these tests can be transfered to system s_2, but the semantics of the tests are not correctly represented by the transfered tests. System s_2 uses multiple instances of the DOCUMENT_LIST Service in one context. Each of these instances represents a subset of all results returned for a query. When one of the above tests is transfered in MTCC, the dialog for the selection of the Service instances addressed in the test is displayed. This approach fails for the tests $t_{1,0}$, $t_{1,1}$, and $t_{1,4}$ because the result list instances in systems s_1 and s_2 are not compatible.

Table 10.8 displays the metrics for question Q3. 88.9% of the tests for system s_4 and 81.1% of the tests for s_3 can be correctly transfered. 60% of the tests for system s_2 can be transfered.

Metric	System	Value
M3.1.1	s_3	81.1%
M3.2.1	s_2	60.0%
M3.3.1	s_4	88.9%

Table 10.8: Metric for question Q3 - transferability of Test Configurations

10.6.1 Validation of the Usability

The usability of MTCC was examined in questions Q4.1 and Q4.2. All domain experts succeeded in the scenario defined by question Q4.1. They were able to construct an initial test from test set TS_{4b} for system s_2 and then use this test as a basis for the definition of the remaining tests from the test set.

Question Q4.2 examines the understandability of the MTCC approach and the usability of the MTCC editor based on 14 tests taken from the test sets TS_1 and TS_{3a}. 52% of all tests could be constructed without help, for further 38%, the domain experts asked for some help but were otherwise able to construct the test.

The validation of the usability and understandability was done with the participation of the domain experts described in Section 10.1 in the time between 2008/03/28 and 2008/05/07. Every domain expert took part in two sessions, in each of which seven tests were modeled. All domain experts worked on the same tests. These tests were selected randomly prior to the validation from the test sets TS_1 and TS_{3a}. The construction of the five tests in test set TS_{4b} examined for question Q4.1 was also conducted during the first session with each participant.

Every domain expert who took part in the validation was given a short introduction to the MTCC editor. The domain experts also had access to a manual describing the basic principles of the editor. The first and the second session were always conducted on separate days, the time between the two sittings was between two days and two weeks.

10.6.1.1 Q4.1 Adaption of Test Configurations

Question Q4.1 examines if the domain experts are able to express the tests $t_{4b,88...90}$ and $t_{4b,103...105}$ using the MTCC editor.

All domain experts were successful in the construction of the tests, on average they took 7.4 Minutes. Table 10.9 presents the time that each domain expert took for the construction and adaptation of the tests.

Participant	Time in Minutes
p_1	5.50
p_2	7.00
p_3	9.00
p_4	8.00
Average	7.38

Table 10.9: Metric M4.1.1 time needed for the adaption of tests

10.6.1.2 Q4.2 Construction of Tests

In order to answer question Q4.2, 14 tests were randomly selected from test set TS_1 and TS_{3a}. For all participants, 51% percent of all tests fall in the category *nohelp*. For 6 tests or 11%, the domain experts were not able to model the tests, these tests were assigned the category *fullhelp*.

The remaining 38% of all tests could be modeled by the domain experts with minimal help, the tests therefore fall in the category *help*.

Table 10.10 informs about the degree to which the participants of the validation were able to model the tests. Some interesting aspects of the validation are pointed out in the following. We discuss and interpret the results in detail in Section 10.7.3.

The percentage of tests in the category *nohelp* is 67% for the second session, compared to 35% for the first session. While the tests selected for the second session cover less test classes than those in the first session, we argue that the experience gained in the first session contributed to the better results in the second session. An interesting point in that context is that all tests that could be modeled by all four participants without help are part of the second session.

Session	Test	*help*	*fullhelp*	*nohelp*
1	$ts_{1,10}$	2	1	1
	$ts_{1,11}$	1	1	2
	$ts_{3a,0}$	2	0	2
	$ts_{3a,10}$	4	0	0
	$ts_{3a,15}$	2	0	2
	$ts_{3a,28}$	2	1	1
	$ts_{3a,5}$	1	1	2
1	Total	14	4	10
	Percentage	50.00%	14.29%	35.71%
2	$ts_{1,15}$	3	0	1
	$ts_{3a,11}$	0	0	4
	$ts_{3a,12}$	0	0	4
	$ts_{3a,14}$	1	0	3
	$ts_{3a,22}$	1	0	3
	$ts_{3a,23}$	2	1	1
	$ts_{3a,8}$	0	1	3
2	Total	7	2	19
	Percentage	25.00%	7.14%	67.86%
1,2	Total	21	6	29
	Percentage	37.50%	10.71%	51.79%

Table 10.10: Metric M4.2.1 Necessary support in test construction

Table 10.11 displays the help needed by the individual domain experts during the validation. All tests in the category *fullhelp* were done by the domain expert p_2. Domain experts p_1 and p_3 did read the manual before each session, domain expert p_4 used the manual during the session, domain expert p_2 did not read the manual.

Person	Category	Count	Percent
p_1	*help*	10	71.43%
	nohelp	4	28.57%
p_2	*help*	5	35.71%
	nohelp	9	64.29%
p_2	*fullhelp*	6	42.86%
	help	4	28.57%
	nohelp	4	28.57%
p_4	*help*	2	14.29%
	nohelp	12	85.71%

Table 10.11: Help needed by the domain experts

10.6.1.3 Problems and Relevant Mistakes in Test Modeling

In addition to the quantitative results discussed above, we also considered qualitative aspects in the validation of the usability. The most relevant cause of problems or faults during the validation was the use of the variables and the operations available for comparing the values of variables. A further source of misunderstandings were the names of Service instances and contexts.

Variables led to the following problems in the validation:

- Domain expert p_3 was repeatedly unsure how and when values were assigned to variables and how variables had to be used. Specifically, the domain expert did not assign values to variables before comparing them with an expected value. Also, often the wrong Test Step was used to assign a value to a variable.

- Domain experts had problems in deciding on the operations used to compare values stored in variables. A particular problem in this context was the decision on the data types that were to be compared. We discuss this point in detail in Section 10.7.3.

- Variables were overwritten on some occasions before they could be used.

An additional problem unrelated to variables was the selection of the correct Test Step instance for a given task. This problem was in part caused by the use of internal variable names for Services and Test Step instances in the editor. The fact that the variable names were in English and were not always correct translations of the German names proved especially problematic.

The correct use of the Test Step FOLLOW_LINK had to be displayed to all domain experts, especially the correct selection of the target Context for a link was only understood after explanation. After such an explanation, all domain experts could use the Test Step successfully.

10.6.1.4 Comments from the Participants

The domain experts who took part in the validation made various remarks during the validation and gave suggestions for improvement. Concerning the MTCC approach itself we found that the representation of a test as a sequence of Test Steps was understood by all domain experts. In addition to comments about MTCC itself, the following feedback was given for the evaluated implementation of the MTCC editor.

- The domain experts liked the limited number of Test Steps and the naming schema for the Test Step instances.
- The domain experts missed the option to copy and paste Test Step instances.
- Changes to the GUI of the editor were suggested, for example, to the controls for adding and removing Test Steps and to the editor for Test Step instance selection.

10.6.2 Validation of the Efficiency

In order to investigate question Q5 of the type III validation as discussed in Section 9.3.6 on page 173, the resources necessary for test automation with MTCC are compared to the resources necessary for manual testing. The time needed for the execution of the manual tests are determined based on the experiences of the domain experts p_2 and p_3. Both participants estimated the time needed for the execution of all tests between 1.5 and 2 hours. Based on a biweekly execution of the tests, 39 to 52 hours a year are needed for test execution. The execution system for the tests in test set TS_{4b} is 124 lines of code long.

10.7 Interpretation of the Validation

Based on the results of the validation, we argue that the MTCC approach as implemented for the purpose of this work is feasible as well as usable and understandable by domain experts.

MTCC models support the representation of 95% of all tests from the test sets $TS_{1...5}$. The MTCC editor is useable with minimal practice and use of the manual. Potential future improvements were identified concerning the usability of the editor and the reuse of Test Configurations for different systems within the system family. In the following, we discuss potential weaknesses in our validation plan and in the conducting of the validation. After this we interpret the results of the validation and derive questions for future work. We also discuss concrete improvement to the MTCC approach.

10.7.1 Robustness of the Validation

The validation of MTCC done for this work is primarily a type I and type II validation [FER08]. Since we chose to do a case study for validation instead of a controlled experiment, the efficiency of MTCC could only be validated in part.

In the following, we discuss the plan and conducting of the MTCC validation. We focus the discussion on potential problems of the validation. We structure the discussion by the criteria defined by Prechelt [Pre01]: reliability, descriptiveness, reproducibility, and generalizability

10.7.1.1 Reliability

The reasons underlying the validation are given in Section 9.2.3. We argue that the validation questions listed are necessary for the validation. We argue that questions Q1, Q2, Q3, and Q4 together with their respective sub questions and metrics are sufficient to support the reliability of the type I and type II validations. As discussed in Section 13.1, the planning and conducting of a type III validation is the subject of future work. Consequently, we do not consider the results of question Q5 as reliable but nevertheless argue that these results are of interest insofar as the allow a first assessment of the efficiency in the absence of more reliable and complete data.

Based on the sufficiency of questions Q1, Q2, Q3, and Q4 concerning the type I and type II validation, we have to examine if the methods and data used to answer the questions are reliable in themselves.

We discussed the motivation for the subquestions and their metrics in Section 9.3.2 and Section 9.3.5. In the following we discuss the test sets and domain experts in terms of their reliability for the validation.

The test set $TS_{1...5}$ used in the validation would endanger the reliability of the validation if they were not representative for the full range of tests that MTCC needs to support. We argue that this is not the case for our validation.

As described in Section 9.3.2, the test sets considered for this validation were collected from various sources and represent requirements from different domain experts. We therefore argue that the test sets are a reliable and sufficient sample of the functional requirements for an automated acceptance test system.

The domain experts who took part in the evaluation and who are discussed in detail in Section 10.1 were all involved in the testing or test design for the member of the system family discussed here. We therefore argue that they form representative if small groups of users for the MTCC approach and specifically the MTCC editor. We consider it advantageous that some of the domain experts took part in the execution of manual tests for the system and thus had prior knowledge of testing. MTCC is an approach to support the automation of tests, not to teach the basics of testing. The fact that the domain experts had skills in manual testing does not endanger our claim that MTCC allows the construction of automated tests by domain experts with little modeling or programming skills.

We argue that four participants and 14 tests are sufficient to judge the basic understandability of MTCC. As Nielsen [Nie94, Nie01] points out, an heuristic evaluation of a system can be done with a limited number of users. The 14 tests used for the validation were selected at random but nevertheless cover the frequently used test classes for MTCC. Optimizations of the usability of the editor and the evaluation of their corresponding implementations must be the subject of future work. We emphasize that we do not consider validation conducted here sufficient to assess the efficiency of MTCC.

As already discussed, such an validation would have to be conducted as a controlled experiment [Mar07] with different control groups.

10.7.1.2 Descriptiveness and Reproducibility

We advance the view that the descriptiveness and reproducibility of the MTCC approach are proven. We describe the derivation of the Domain Feature Model and Domain Test Model for the considered system family and the construction of the Application State Model and Application Feature Model for systems $s_{1...4}$ in Chapter 8. Excerpts from the test sets used for the validation can be found in Section A of the appendix.

Systems s_1, s_2 and s_3 are publicly accessible, only s_4 is an internal system. All the software used for the implementation is available as Open Source.

The validation of MTCC is described in detail, the details of the approach are available and can be used to replicate and verify the validation.

10.7.1.3 Generalizability

When discussing the generalizability of the validation of MTCC it is important to consider the fact that an application of MTCC is by definition specific to a system family and can only be validated in the context of this system family. It could therefore be argued that the results of this validation only apply to the system family of Digital Libraries discussed here.

We advance the view that the results of the validation can be generalized insofar as no reasons exist that would hinder the adaptation of MTCC to other dialog oriented systems besides Digital Libraries.

10.7.2 Feasibility of MTCC

We argue that the practicality of MTCC is proven by the results of the type I validation that was conducted based on the questions Q1, Q2, and Q3.

95% of all tests considered in the validation could be represented by MTCC. The respective Test Configurations could be used for the automatic generation of test code. Opportunistic reuse of Test Configurations for the different system models is possible for more then half of the Test Configurations in the validation.

10.7.2.1 Interpretation of the Expressiveness

Table 10.2 displays the percentage of tests from the test sets $TS_{1...5}$ supported by MTCC. If entries from the test sets that fall in the category $\mathrm{g}_{\mathrm{extern}}$ are ignored, 94.84% of all tests are supported in MTCC.

The tests that are not supported are mostly from the test sets T_{4a} and T_5. Tests from these test sets that fall in the category g_{no} are mostly members of two groups of tests: (1) Tests whose execution is not supported by MTCC since external applications would have to be exercised and (2) tests that address specific aspects of the GUI of the testee that are abstracted in MTCC.

The following tests can be expressed by MTCC Test Configurations, but fall into the category g_{no} since they cannot be executed by MTCC:

- Test $t_{3a,18}$ requires access to the Refworks[4] application.
- Test $t_{5,184}$ verifies the correctness of the export of result documents by email.

The following tests are examples for entries from test set TS_5 that cannot be expressed in MTCC since they are dependent on specifics of the GUI of the testee which is expressly abstracted away in MTCC.

- Test $t_{5,31}$ exercises the Drop Down list that is used to change the order in which documents are displayed in the result list. It checks whether the Drop Down List contains the correct values in the correct order. Since MTCC does not model the specifics of the user interface, this test cannot be represented.
- Test $t_{5,281}$ is adapted from a fault reported for the GUI. One of two instances of a control used for browsing a list of results was not operational. Again, MTCC is not able to represent the test since the specifics of the GUI are not represented in the MTCC models.

The metrics for question Q1 in Table 10.3 summarize the results of the validation for question Q1.

MTCC provided support of 89.13% for the test set TS_{3b}. Two of the tests in the test set are not supported because the execution of the tests would require the execution of external Services.

We argue that the feasibility of MTCC is not challenged by the fact that MTCC does not support such tests. The automation of tests that rely on the interaction with external Services is also problematic for other test automation approaches. The testing of GUI functionality is not the purpose of MTCC and can be better tested by dedicated approaches for GUI testing.

10.7.2.2 Execution of Tests

We consider the executability of MTCC to be proven by the results of the validation. The implementation of the execution system was possible with limited resources, generated tests could be executed by a COTS test system. As discussed in Section 10.5, the execution systems consist of 925 lines of code for all implemented Test Adapters and 596 lines of code for testing core assets. We argue that this resource usage is acceptable given the number of tests that can be expressed based on the implemented infrastructure, especially when the fact is considered that the code of the Test Adapters is structurally simple. The feasibility of the test

[4] http://refworks.com

code generation approach could be proven. The MTCC implementation used for this validation generates tests for the Python xUnit implementation.

We expect an implementation of an execution system for other programming languages to be unproblematic. The abstractions used by the Test Parameter objects provide common abstractions that are provided by most programming languages and no assumption about the execution of tests is made.

10.7.2.3 Reuse

The reuse of Test Configurations for different system models is possible for 50% of all tests of test set TS_1 and the target systems s_2, s_3, and s_4. The tests that are not reusable mostly fall in the category g_{no}. This category consists of tests that cannot be reused because of missing or otherwise incompatible features of systems. Reuse is therefore not possible regardless of the test reuse approach used.

As discussed in Section 10.6, the test $t_{1,0}$, $t_{1,1}$ and $t_{1,4}$ can be reused for system s_2, but do not correctly represent the intent of the test. We therefore consider these tests as not automatically reusable.

The test $t_{1,2}$ is in the category $g_{can}$for all systems in the system family S — the test is reusable in theory but cannot be transfered by the MTCC reuse system. The reason for this failure of the automatic reuse system is a lack of support for mappings between different names and representations for the same concept in the feature models. Specifically, a filter in s_1 could be represented as a semantically equivalent search field value in s_2, but an automated mapping of this concepts is not supported.

Between 60% and 90 % of all tests for a system can be automatically reused if only those tests are considered in the validation that do not fall in category g_{no}. We argue therefore that the support for the reuse of tests that MTCC facilitates already allows for significant savings in testing costs. Based on this, we further argue that reuse in MTCC is successfully implemented.

10.7.3 Usability of MTCC

Based on the result of the validation of the usability of MTCC, we argue that the approach is understandable and usable given minimal training. While significant potential for improvement exists regarding the usability of the editor, the prototype implementation used in the evaluation can be successfully used for the construction of tests.

As Table 10.10 displays, 52% of the tests could be modeled without help, for 37% some help was needed and for 11% the tests could not be modeled. Since the same domain expert was responsible for all failed tests and because this domain expert did not read the manual, we argue that MTCC can be understood with minimal preparation. We also advance this view especially given the fact that the results of the second session — 68% of all tests could be modeled without help — were significantly better than the first session.

As we discussed in Section 9.3.5, one re-occurring problem in the use of the editor was the use of variables. We discuss potential solutions for this problem in Section 13.1.

As we pointed out in Section 10.6.1.4, the participants of the validation made a number of suggestions to improve the usability of MTCC. We argue that these improvements can be addressed in separation from the other issues discussed in the validation and that the MTCC approach is valid and usable and that the usability can be improved in the future even though the usability of the editor was not the focus of this work [Nor06].

10.7.4 Efficiency of MTCC

Based on the limited validation of the efficiency of MTCC, the automation of the execution of test set TS_{4b}, we argue that the use of MTCC is beneficial from a resource point of view even over short periods of time. A robust validation of the efficiency, however, is the subject of future work.

As discussed in Section 10.6.2, 39 to 52 hours of time per year are spent on manual test automation. The implementation of the execution system for the tests in test set TS_{4b} consists of 124 lines of Python code.

If 4 hours are estimated for the implementation and maintenance of the execution system, which we argue is realistic based on 124 lines of code and the simple structure of the test code, and if 15 minutes are needed for each run of the automated test, then testing with MTCC requires 10.5 hours per year versus a minimum of 39 hours for manual testing. If we further consider the fact that the addition of new tests to the test set TS_{4b} would have no impact on the resources required by MTCC, then MTCC is significantly more efficient than manual testing, even for a limited number of tests and one system under test.

10.8 Chapter Summary

The validation of MTCC was conducted on three Digital Libraries and the installation of the Solr search server at the GESIS. Four domain experts were involved in the validation of MTCC. All domain experts had prior experience in the specification or execution of tests or information retrieval evaluation scenarios. Five different sets of tests were used for the validation of MTCC. These test sets include tests taken from the testing processes of the evaluated systems as well as tests that were contributed by domain experts for the validation and tests derived from entries in an issue tracking system.

The type I validation showed that 95% of all tests from the investigated test sets that were of interest to the MTCC approach could be expressed by MTCC test models. The execution of tests was successful for the validated test set and for all considered SUTs and test runners. The implementation of the execution system was done in 925 lines of Python code, an implementation of all 23 Test Steps needed for the execution of all tests in test sets TS_1 and TS_3b was done in 530 lines of code. Between 60% and 88% of all practically transferable tests for system s_1 defined in the test set TS_1 could also be transfered by the MTCC test reuse system.

The type II validation of MTCC was conducted in eight sessions. In each of these sessions, the four domain experts modeled seven tests. Out of the 56 tests modeled in total, the domain experts could represent 52% without any help, a further 37% of the tests could be expressed with minimal help that would also have been available from the MTCC manual. 11% of the tests could not be expressed.

We conclude that the MTCC approach is feasible, can be implemented with limited resources and can be understood by domain experts given minimal training.

11 Related Work

This chapter examines approaches that address situations similar to those considered by MTCC, are faced with related challenges or use comparable methods to tackle the problem of abstracting testees that have to be exercised by automatic tests.

One basic difference between MTCC and approaches to model-driven testing [BBN04, BBN02] is that MTCC is not based on the principle to automatically generate tests from a model. Specifically, MTCC does not aim to reuse models from the software construction process. Bertolino describes the motivation behind such reuse in the following: *"The leading idea is to use models defined in software construction to drive the testing process, in particular to automatically generate the test cases."* [Ber07]

Models in MTCC, in contrast, are used to abstract implementation details and facilitate the systematic handling of the variability of a system family. The goal of MTCC is the construction of tests by domain experts, not the automatic creation of tests.

Given these fundamental differences — regarding both the goals and the methodology — between MTCC and automatic, model-driven testing, we do not discuss model-driven testing aimed at fully automatic test generation further.

This chapter discusses different fields of research related to MTCC and gives examples for concrete approaches. Section 11.1 investigates the use of use cases and examples for testing. Section 11.2 discusses approaches that employ model-driven software development techniques for the realization of test code. Section 11.3 introduces approaches that represent graphical user interfaces and web applications as models for testing. Section 11.4 examines approaches to software product-line testing, in particular an environment for test construction that is in some aspects comparable to the MTCC editor. Section 11.5 discusses approaches to involve users and domain experts in the programming or more general software development.

11.1 Testing Based on Usage Scenarios

We call approaches that aim to capture and formalize requirements and use them as a basis for testing and test generation requirements-based testing. Approaches to requirements-based testing frequently build on the notations and methods of the requirements engineering process and then extend and leverage them for testing.

The most significant difference between approaches to requirements-based testing and MTCC — beside the focus of the former on the testing of single systems and the aim to reuse and extend methods and artifacts from the software construction process — is test construction by domain experts as a central aspect of MTCC.

One way to express requirements are scenarios, defined by Ryser as follows: *'Descriptions of interaction sequences between two or more partners, most often between a system and its users'* [Rys03]. A use case is an extension of a scenario that also describes alternate scenarios for special cases or failures.

A central aspect of requirements based approaches is the formalization of existing notations of requirements [TBPY01, Rys03, Pos96, NF06, Pos98] in order to turn them into testable [Bin99] models.

MTCC considers artifacts for requirements analysis and testing in separation. While MTCC Test Configurations serve as a limited, functional, executable specification [HS00, Fow06], they are intended to supplement, not replace, existing specification.

One group of approaches to requirements-based testing is based on UML [GEMT06, BPR03], primarily on use case diagrams. Regarding their relationship to MTCC, these approaches are no different from the approaches discussed above.

11.1.1 The SCENT Approach

The SCENT approach developed in Ryser [Rys03] addresses the verification of requirements and the conducting of system-level tests based on scenarios expressed as state-charts. A particular focus of Ryser's work are dependencies between state-charts. Domain experts are involved in the identification and design of scenarios and thereby contribute to the construction of high-quality software systems. SCENT differs from MTCC in the consideration of single systems and in the way in which domain experts are involved in the testing process. The SCENT approach is subdivided in five steps:

- Use cases for the considered system are captured as SCENT scenarios. The analysis of the testee for scenarios is done by software engineers and domain experts in cooperation. The goal of the approach is the description of scenarios in natural language.
- Every scenario is modeled as a state-chart that represents every system event as a state and every user interaction as an event.

- Tests are derived from the models of scenarios and their dependencies. This derivation will in most cases be done manually.
- The state-charts are enhanced with information that is later used for the generation of tests from the model. Among other information, pre- and post-conditions are added to the states of the state-charts. Also, supported ranges for input values are added for the user events.
- The dependencies among the previously defined state-charts are analyzed and formalized. This is a prerequisite for the later composition of scenarios into tests. SCENT uses a graph-based notation to express sequences, alternatives and relations for scenarios. Figure 11.1 gives an overview of some of the relevant concepts.

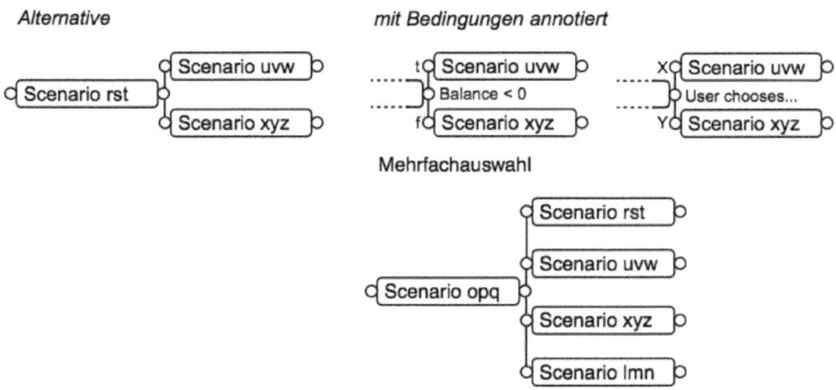

Figure 11.1: Notation to express dependencies of scenarios in SCENT [Rys03]

SCENT is similar to MTCC both concerning its realization as well as its goals. Both MTCC and SCENT envision the involvement of domain experts in testing and consider testing as an essential activity during the software lifecycle.

Basis differences between MTCC and SCENT exist in the concrete roles that domain experts have in both approaches, in the test construction process and in the treatment of system families in MTCC. These differences result in different requirements for the models used in the approaches.

MTCC involves domain experts during the analysis of a system for its test relevant aspects as well as during the construction of tests. SCENT involves domain experts in requirements analysis and in the realization of test scenarios. These scenarios are not tests in themselves but serve as an input for test derivation.

The central artifacts in SCENT are not tests but the formalized scenarios and the dependencies between these scenarios. In contrast, the focus of MTCC is the construction of concrete tests. We argue that the MTCC approach is better suited to facilitate the maintenance of tests during the software development process. In MTCC, domain experts can construct tests at any given time without the need to create executable tests from scenarios. Since SCENT envisions the automatic generation of tests from the scenarios, these scenarios have to contain more information about the testee than needed in MTCC. While this causes an increase in the complexity of the models in SCENT and makes modeling more expensive, it can potentially result in a better test coverage than testing using the MTCC approach. Both approaches depend on the cooperation of domain experts and software engineers during requirements analysis.

MTCC and SCENT use state-charts or finite state machines to represent the behavior of systems under test, specifically to represent possible interaction sequences between users and the system under test. The approaches differ in the granularity in which systems are modeled — MTCC only includes high-level models of the behavior of a system.

One significant difference between MTCC and SCENT is the consideration of system families by MTCC.

11.1.2 The Eg Approach

The Eg approach developed by Gaelli [Gäl06] uses scenarios, there called examples, for the testing of software as well as for documentation. This approach is related to MTCC insofar as Gaelli does not aim at the reuse of artifacts from design or requirements analysis.

Important differences between the Eg approach and MTCC exist both in the fact that Eg addresses software engineers and not domain experts and in the use of tests as examples, not as specification. Figure 11.2 illustrates this difference between MTCC and Eg: The acceptance tests constructed with MTCC correspond to story tests in the figure, Gaelli's approach addresses unit tests, the user of the Eg approach are developers. As MTCC, Eg does not restrict the role of tests on fault finding, but where MTCC considers tests as a executable specification, Eg considers them as examples that facilitate the understanding of software.

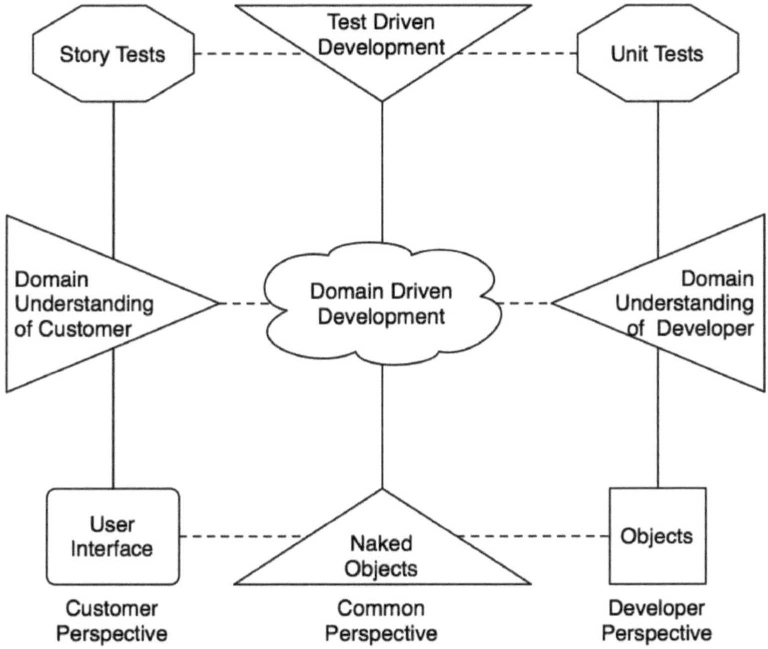

Figure 11.2: Viewpoints of domain experts (customers) and domain engineers (developers) [Gäl06]

Figure 11.3 displays the Eg meta model. As the figure illustrates, Eg addresses a system under test on the class level. This fine-grained view is necessary to support the implementation-based tests envisioned by Eg, but is a basic difference to the abstract MTCC feature models that explicitly abstract the implementation.

11.2 Modeling of Tests

Most approaches to test modeling do not aim to derive or generate tests from models but rather use a special notation to construct or model tests directly. The models or notations used in such approaches are dedicated to test construction, existing models from software design are not reused.

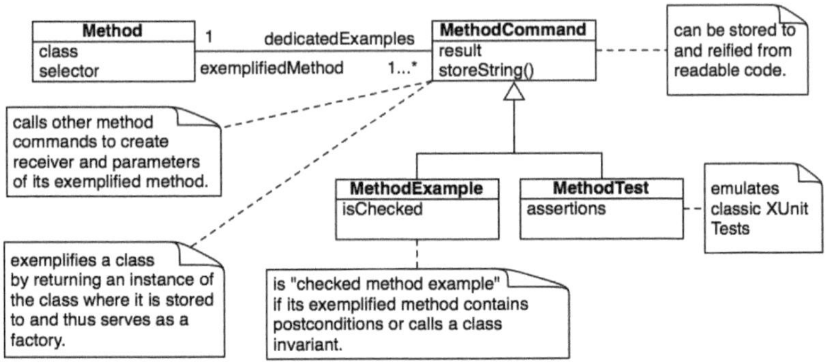

Figure 11.3: The Eg meta model [Gäl06]

One difference between the approaches discussed here and MTCC is that the notation used for test description in the former is less abstract than the models used in MTCC. For example, tests are expressed in an XML notation or created using a spreadsheet application [And04b, And04a]. One example for the construction of tests as tables is FIT [MC05a, MC05b, Mug04]. Like MTCC, FIT is an approach for the automation of acceptance tests. We discuss the relationship of FIT and MTCC in Section 11.5.

11.2.1 UML 2.0 Testing Profile

The UML 2.0 Testing Profile [SDGR03, BDG$^+$03] adds concepts relevant for testing to the UML. The application of the UML2TP allows the modeling of tests. The automatic generation of test code from the models is envisioned [Dai04]. In contrast to MTCC, the U2TP is not a complete approach to testing but rather a specialized profile for test modeling.

A well established way to express tests independent from the specifics of a GUI are keywords and action words [GF00]. Like MTCC, approaches based on keywords or action words identify relevant actions on the system under test and abstract the specifics of the GUI [AB04]. Each keyword or action word represents a concrete step in a workflow or in the interaction between the testee and a user. Keywords and action words correspond to Test Steps in MTCC, but a difference exists in the fact that MTCC builds on this common concepts by providing a model based editor and considering system families instead of single systems.

Test scripts are no different from other software in the fact that they can be generated in a model-driven software development process. One example for an approach that uses model-driven development in the testing domain can be found in [SKS06].

11.2.2 Model-Driven Test Development

Approaches for model-driven software development can generate test code as well as application code. One approach that applies MDSD techniques to the testing domain is the work by Sensler et al [SKS06]. Building on an initial analysis, tests for a system are formalized in XML, this XML then serves as input for a code generation system. With the consideration of system families and the goal to include domain experts in the testing process, MTCC addresses goals beyond the separation from test models and specific system implementations addressed by such systems.

11.3 Abstract Representation of User-Interfaces

MTCC is an approach that abstracts user interfaces in order to allow the description of high-level function tests. Therefore, MTCC follows a different approach than approaches that seek to test the GUI itself.

11.3.1 Representation of GUIs

Memon [MPS01, Mem01] uses a hierarchical model to describe the possible interactions of a user and a GUI. Tests are created automatically based on a test plan and the model of the GUI. Like in MTCC, structural and dynamic aspects of a testee are modeled, but differences exist in the granularity in which the GUI is represented. While MTCC only represents the GUI in terms of the available Test Steps, Memon's representation also includes GUI elements like menu items and other low-level artifacts.

Li et al [LHRM07] use a two-level model for a GUI. An interesting aspect of their work is the GUI component that roughly corresponds to a Service in MTCC. A fundamental difference between MTCC and the approaches above is that the models used in MTCC do not represent the GUI, but the functionality provided by the GUI. This allows the use of a common domain model for systems with fundamentally different GUIs.

11.3.2 Modeling of Web Applications

Ricca [RT01, Ric03] developed the Reweb approach that models the structure of a web site as a graph structure. The representation of a web site as a graph structure is similar to the representation of a testee in the MTCC Application State Model.

A relevant difference between approaches like Reweb and MTCC is the fact that MTCC, while it is validated in the web application domain, is technology agnostic. Reweb in contrast is specific not only for the behavior and structure typical for web application but also for concrete implementation technologies as can be seen from the Reweb meta model displayed in Figure 11.4.

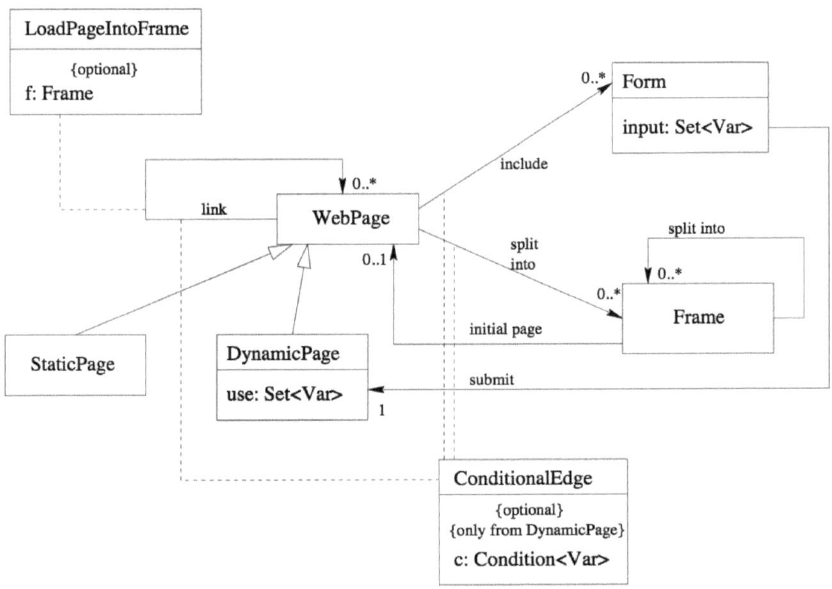

Figure 11.4: Meta model of a web application underlying the Reweb approach [RT01]

11.4 Testing of System Families

One central aspect of MTCC is the testing of system families, MTCC is therefore related to approaches for system product-line testing and to alternative approaches that address the testing of system families without systematic reuse. Fundamental differences between MTCC and these approaches exist both in the underlying process and in the view of the respective testees.

11.4.1 Testing of Software Product Lines

A central question concerning the testing of products from software product lines [TTK04] is the selection of those software systems from all possible systems of the product-line that are to be tested [GLRW04]. Another, even more basic issue is the capture and treatment of the variability in the testing context [BG03b, BG03a].

The problem of test selection is not relevant in the context of MTCC, because tests are selected by domain experts. The treatment of the variability of the system is not done on the level of individual tests but is done on the level of the system as expressed by the Application State Model and the Application Feature Model. In MTCC tests are constructed for one concrete member of a system family.

The MTCC editor can be regarded as a tool for the manipulation of feature diagrams. It differs from other editing environments [MS05, AC04] for feature models as displayed in Figure 11.5 insofar that its goal is not the representation of a feature model and supporting specializations on this feature model. Instead, the MTCC editor aims to offer a view of the feature model that allows domain experts the configuration of tests based on domain-specific GUI conventions.

11.4.2 Condron's TADE Approach

The TADE (Test Application Development Environment) [Con04] presented by Condron has a number of similarities to MTCC: The approach uses an editor to support the specification of tests for the members of a systen family. Figure 11.6 displays a screen shot of the TADE.

The most relevant differences between Condron's approach and MTCC are regarding the goals of the approach, the use of models and the reuse of test models. The goal of Condron's approach is to decouple tests form implementation details of individual testees and reuse the testing infrastructure. MTCC aims at the reuse of tests and the involvement of domain experts. Regarding the use of models, the domain and systems models, their construction, specialization and composition are central aspects of the MTCC process. This use of models allows the systematic treatment of the common and variable test-relevant features of the member of a system family.

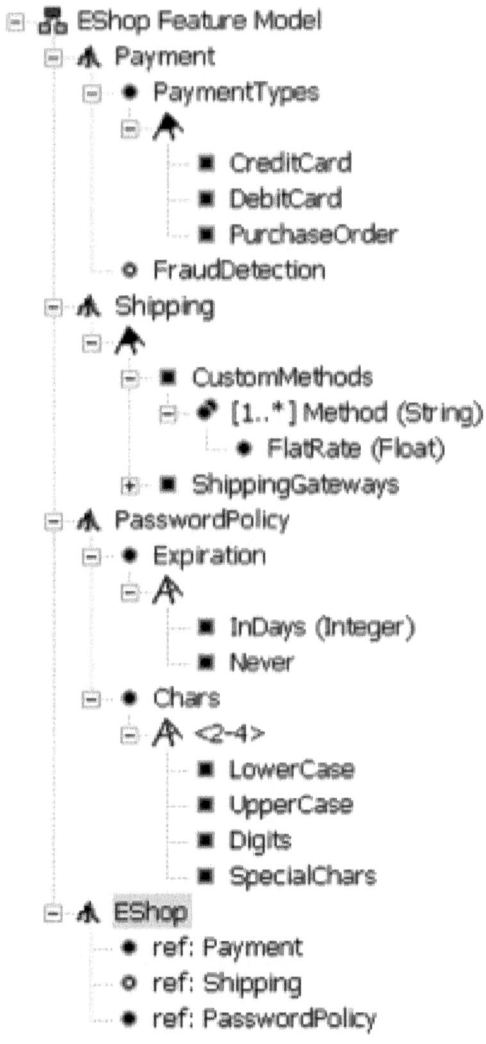

Figure 11.5: Representation of a feature model in an editor [AC04]

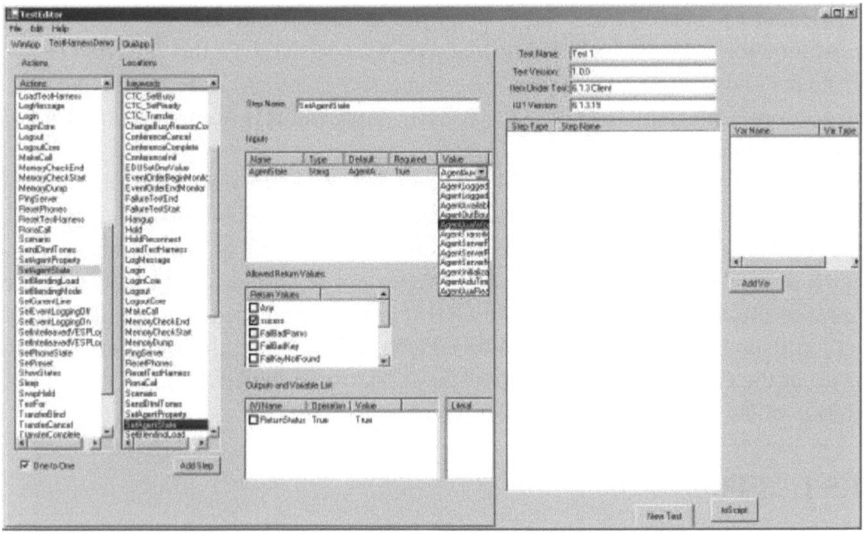

Figure 11.6: GUI of the TADE for test development [Con04]

The TADE approach presented by Condron is not model-based or model-driven in the sense used in the work. Finally, Condron does not address the reuse of test models for different variants or versions of a system.

11.5 Involvement of Domain Experts in Software Development

The automatic instantiation of graphical user interfaces for modeling systems usable by end users is an area of active research. We differentiate generic approaches without a specific application domain [PNT06] from those that are designed for a specific purpose and for a concrete group of users, for example, for the development of component-oriented systems [Ufl05] or the programming of machine tools [PHM06]. Examples for systems that are not specific for an application domain are meta case systems [TR03, DFG06]. All the systems above aim to involve people without formal modeling skills, but differ in the fact that they do not address the testing domain.

An example for a commercial system that facilitates end-user programming is the Automator software included in Mac OS X[1]. Like MTCC, this program represents a program or test as a sequence of successive steps and supports the parameterization of each step. In contrast to MTCC, Automator is not based on a formal GUI-independent model.

An alternative to the use of a GUI like in MTCC is the use of simple, textual domain-specific language for end user programming [Spi01, vDKV00].

MTCC differs from the above approaches insofar as it is a model-based approach for testing. The models used to represent tests and testable systems are central to the approach, the concrete GUI representation used in the editor is decoupled from the models but could be adapted by changing the Buildlets used to construct the GUI. The relationship of the models and the GUI can be thought of in terms of the model-view-controller approach [BMR+96].

11.5.1 FIT

FIT, like MTCC, is an approach to the automation of acceptance tests. MTCC and FIT have a number of similarities regarding their respective goals as well as aspects of their realization.

A basic difference between FIT and MTCC is the consideration of system families in MTCC and the resulting differences in the underlying models that separate Test Configuration and test execution for MTCC. FIT does not address the systematic treatment of the variability within a family of testees.

Like MTCC, FIT considers the role of testing to be both in fault detection and in the identification and clarification of requirements. Tests are derived in story-driven test development, relevant use cases for testing are identified and serve as the basis for tests and the identification of the parameters needed for testing.

FIT supports different types of tests. Generally, tests either verify the correctness of calculations or exercise a testee based on a sequence of action words. The later type of tests is called a workflow test. All tests in FIT are expressed as tables. This representation was chosen because tables and spreadsheets as well as the tools used to interact with them are familiar to non-technical users.

Figure 11.7 displays a sketch of a table for a FIT workflow test. Such a sketch would be used in the analysis phase to identify functionality relevant to testing.

FIT uses fixtures to execute a test described in a table on the testee. Fixtures in FIT broadly correspond to Test Adapters in MTCC — they provide an implementation for the actions specified in a test. Figure 11.8 illustrates the relationship

[1] http://developer.apple.com/macosx/automator.html

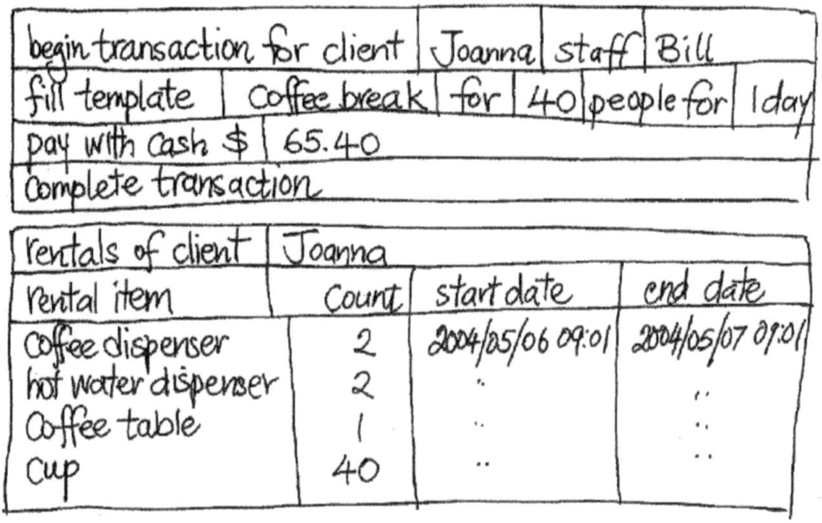

Figure 11.7: Sketch of a FIT workflow test [MC05b]

between tables and fixtures. In contrast to MTCC, the GUI representation of a test and the underlying model of the test are not different concepts in FIT. The table used to represent the test is both the GUI and the model.

FIT does not address the testing of different systems in a system family. While it is possible to implement multiple fixtures to execute a test on different systems, such an application is not an explicit part of FIT.

11.6 Novelty of MTCC

MTCC is novel in its goal to facilitate the involvement of domain experts in model-based testing of software system families. In order to achieve this goal, MTCC combines the fields of acceptance testing, model-driven software development, feature modeling and software product-line engineering. The main attributes that distinguish MTCC from existing approaches are (1) the use of a hierarchy of models to support the intellectual specification of tests and facilitate the reuse and maintenance of these tests. (2) The use of an editor to represent this models to domain experts as a GUI based on the conventions of the domain.

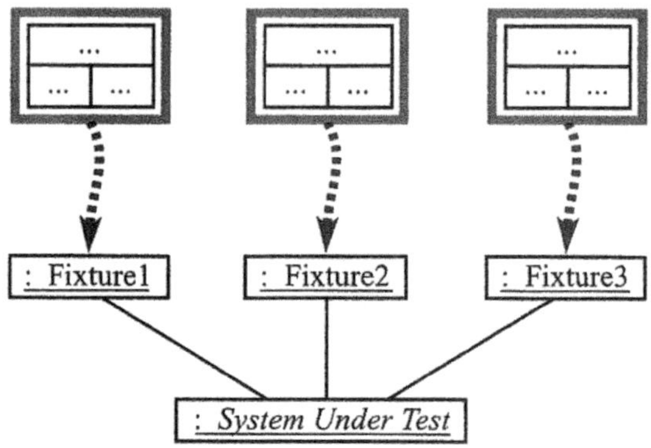

Figure 11.8: Relationship of testees, tests and fixtures in FIT [MC05b]

11.7 Chapter Summary

MTCC is related to a number of different approaches and research directions in Software Engineering. MTCC differs from these approaches in its goals as well as in the research directions taken to address the various challenges and research questions.

Requirements-based testing reuses and enhances existing requirements representation as a basis for testing. MTCC does not assume that formal requirements exist and uses dedicated models to represent the SUT and test relevant for the SUT. The use of models specific to testing also puts MTCC in contrast to approaches like the U2TP that aim for the reuse of design models for testing.

MTCC is not an approach to model-driven GUI testing. The models used by MTCC represent the Services that are made available through the GUI, not the GUI itself. MTCC differs from approaches to software product line testing in that it does not assume the existence of a common set of Core Assets.

A central difference between MTCC and the above approaches is the strong focus that MTCC puts on the involvement of domain experts in the testing process. MTCC differs from alternative approaches to automated acceptance testing in the use of models to systematically describe the available features in a system family and the Test Steps available to exercise these features. The combination of models dedicated to testing members of system families and the involvement of domain experts, facilitated by the MTCC editor, are novel contributions of MTCC.

12 Summary and Contributions

This work introduced and validated the MTCC approach to model-driven acceptance testing for system families. MTCC applies methods from software product line engineering to the testing domain and uses models to represent the test-relevant features of a system. Tests represented by models and serve as the basis for test code generation. MTCC makes several contributions to the domains of automated acceptance testing and to the testing of system families.

12.1 Summary

The MTCC approach introduced in this work supports the construction of acceptance tests for the members of a system family by domain experts. In this work, MTCC was applied to a Digital Library system family. The test-relevant functionality and workflow steps of the systems in the system family were captured in MTCC models. These MTCC models and an implementation of the MTCC core assets serve as the basis for the validation of MTCC.

The feasibility and practicality of MTCC as well as the usability of the MTCC editor were evaluated with good results.

The basic hypothesis of this work is that the use of models supports the construction of abstract tests by domain experts using an editor. We further advanced the hypothesis that the use of abstract models facilitates the reuse of tests in a system family context. We argued that abstract test models can be leveraged to exercise and test concrete systems. We introduced the MTCC process and its artifacts in Chapter 5 in order to demonstrate the validity of the approach.

12.1.1 The MTCC Approach

The MTCC process considers the representation of systems and Test Steps on three levels. On the domain or system family level (1), members of the family of testees are analyzed for commonalities and variability regarding their test-relevant functionality. The findings of this analysis are formalized as Services and Test Steps.

On the system level (2), a representation for a concrete system is built by specializing the feature models that represent Services at the domain level to concrete instance representations at the system family level. On the testing level (3), the models that represent the behavior, features, and tests of MTCC at the system and domain-level are integrated into a common model that is dedicated to test construction. This model represents a system in terms of the possible sequences of configurable Test Steps that can be executed on the system.

The actual construction of tests is done using the MTCC editor. The editor represents the MTCC models according to the conventions of the application domain and allows their configuration. Constructed tests — test configurations — are used to generate test cases that are executable with COTS test runners.

12.1.2 Applying MTCC to a System Family of Digital Libraries

We have applied MTCC to a system family of three Digital Libraries and on the search server Solr. The identification of relevant tests was done based on an already established testing regime, documented faults and requirements by domain experts. The identified tests include scenarios for the evaluation of Information Retrieval systems as well as functional tests.

12.1.3 Implementation

We implemented the MTCC core assets, specifically the editor, the domain model and the Test Generator for this MTCC evaluation. Test Adapters were implemented to different degrees for all four systems considered in the validation and for two COTS test runners.

12.1.4 Validation

A type I and type II validation for MTCC was done based on the GQM approach. The type I validation assessed the feasibility of the MTCC approach. The capability of MTCC to represent the relevant tests for the domain was validated. Additionally, the reusability of test models between different system models and the executability of MTCC models for the testing of system implementations were proven. A type II validation was conducted with the involvement of four domain experts. The validation lead to the following results:

- 94.84% of all tests considered in the validation could be expressed with MTCC. Tests that are not supported by MTCC are either dependent on specifics of the testee's GUI or exercise external Services.

- The Test Adapter instances needed for the validation could be implemented with few resources, specifically 900 lines of code.
- The reuse of Test Configurations for different members of the examined system family is possible for about 50% of all tests. The reuse techniques are limited by the fact that MTCC can not verify that tests that were transfered to a new system are semantically equivalent to the original tests.
- The MTCC approach is understandable and usable by domain experts given minimal training and preparation.

12.2 Contributions

We see the scientific contribution of MTCC primarily in the modeling concepts developed to express the testable features and tests for systems in the context of product-lines and in the concepts used to represent and manipulate the models as they are used in the editor. Specific results from the application of MTCC to the domain of Digital Libraries in this context are the domain-level models that capture the aspects of the domain relevant to acceptance tests and the system-level models that represent individual systems. The usage of feature models to describe tests in the context of a system family and the transfer of specializations that were applied to these feature models in order to facilitate test reuse are — to our knowledge — original contributions of MTCC.

Specifically, MTCC makes the following contributions to automated testing

- The process for test construction used by MTCC describes how domain analysis and software product-line methodologies can be applied to testing. With the identification of test-relevant functionality and actions, the formalization of this knowledge in domain-level models, and the specialization of these models to system-level models, MTCC defines a systematic process to realize a testing infrastructure for an existing system family.
- One novel aspect of MTCC is the use of feature models to represent the testable features of the members of a system family. MTCC employes feature models for the description of high-level Services as well as for the definition of parameters that are available for individual Test Steps. The separation of system-level Service instances and domain-level Test Steps, in combination with the use of references between feature models, reduces the redundancy in test modeling and thus supports maintenance and adaptation of test models.

- The systematic reuse of existing Test Configurations is an original contribution of MTCC. By checking the transferability of Test Step sequences both based on the features referenced by individual Test Step instances and on the sequence of Test Step types, MTCC enables the reuse of existing tests.

- On significant contribution of MTCC is the realization of an approach that allows the representation of feature models in an editor for tests as well as the manipulation of these feature models. The combination of Buildlets and configuration nodes allows for the representation of feature models according to the standards and customs of the application domain investigated by MTCC.

In addition to the points above, MTCC also made the following contributions:

- In the course of this work we applied MTCC to the Digital Library domain. This work resulted in MTCC domain and system level models for the test-relevant functionality of a system family of Digital Libraries. Since these models are based on the requirements of domain experts and on current testing practice for the considered systems, the test models represent the functionality that has to be covered by testing approaches within this domain.

- The validation of the MTCC approach outlines possible validation scenarios for other approaches with similar goals.

- The implementation of the MTCC execution system indicates the resources necessary to implement a test library for a web application from the Digital Library domain.

- The validation conducted as part of this work demonstrated that domain experts using the MTCC approach can contribute to automated testing given minimal training.

13 Outlook and Conclusion

While the feasibility and the practicality of the MTCC approach were demonstrated by the MTCC validation, a number of interesting questions remain future work

13.1 Future Work

We differentiate two major areas of future research for the MTCC: The further validation of the approach and its extension and adaptation based on the results of this work.

13.1.1 Additional Validation

Three topics are relevant for future validations: The feasibility of MTCC for other application domains other than Digital Libraries, the assessment of MTCC in the field and the conducting of a type III validation:

- This work validates MTCC for a system family of digital libraries. In order to assess the feasibility and practicality of MTCC in other application domains, the capability of MTCC to represent the concepts of these domains must also be assessed. Beside the conceptual aspects, the ability of MTCC to exercise testees that are not web applications must also be verified.

- While MTCC was validated on production systems, the approach was not yet used in a production context. One question that a future validation could address is to what degree MTCC can replace or supplement an existing testing approach. An interesting question in this context is whether MTCC is primarily useful in the context of regression tests or if it also supports the identification of previously unknown faults.

- In the context of this work, a type III validation was only done to a very limited degree. In order to compare the efficiency of MTCC to other approaches to acceptance test automation, a controlled experiment that compares MTCC to FIT would be of interest.

13.1.2 Additions to the MTCC Approach

Two directions are possible for future work on the MTCC approach: The addition of new Test Steps and the improvement of existing tooling and changes to the fundamental aspects of the approach:

- The addition of new Test Steps or Services to the MTCC approach is the least complicated way to allow the testing of new functionality with MTCC.
- For this work we created the domain and system-level models used by MTCC programatically. The implementation of the tooling to create and edit MTCC models would facilitate the direct involvement of domain experts in the domain engineering and test engineering stages.
- MTCC does not provide a type system for attribute values. More generally, MTCC does not support constraints that would allow to distinguish valid and invalid feature configurations. The expansion of MTCC by a type system or constraints would make it possible to detect and display invalid Test Configurations in the editor.
- Test execution in MTCC is done offline, tests cannot be executed while they are constructed in the editor. If the capability of online test execution was added to MTCC, domain experts could verify that a test behaves as expected during test execution.
- In their present form, a sequence of Test Steps can only be transfered unchanged to a target system. Future work would have to investigate the transfer of tests if an existing sequence of Test Steps cannot be transferred but a semantically equivalent sequence exists.
- One way to guide domain experts in test construction would be this display of the test coverage for a specific Service of context of the SUT.
- A future subject of research could be the utilization of MTCC models to test the conformance of GUI with the WOB model [Kra06b] by defining a suitable domain model that represents the attributes of the WOB model and than test the compatibility of systems with this model.

13.2 Concluding Remarks

MTCC demonstrates that model-driven automated acceptance testing can be realized for system families and that such testing is usable by domain experts. The

MTCC approach builds on concepts from model-driven development and software product-line engineering to allow the involvement of domain experts in a testing process that facilitates the systematic treatment of the variability and commonalities in a system family. MTCC allows automated testing based on models for application domains that are characterized by vague, user-defined requirements which are hard to formalize for fully automatic model-driven testing approaches that do not involve domain experts.

In addition to further work on the MTCC approach and its validation, we envision the use of MTCC to test the systems presented in the validation and thereby both increase their quality and save human effort that would otherwise be necessary for test execution.

A Test Sets

Test Set TS_1

Listing A.1: Excerpt from TS_1 : Tests defined for the MTCC validation

```
<?oxygen RNGSchema="testsets.rnc" type="compact"?>
<out system="sowiport">
    <div category="yes" id="ts_1_0" scenario="ts_1_0.xml">
        <!— Kein Titelfeld in ü—>
        <reuse system="iblk" category="no">Kein Titelfeld im IBLK</reuse>
        <reuse system="solr" category="yes"/>
        <!— Technisch übertragbar, semantisch problematisch wegen der
    verschiedene, getrennten Ergebnislisten—>
        <reuse system="infoconnex" category="yes"/>
        <h2>Test 1(bs) Suche mit Umlaut</h2>
        <p>Dieser Test überprüft die Suche mit dem Umlaut ü</p>
        <ol>
            <li>Aktion In der Suche wird im Feld Titel nach Überleben
         gesucht</li>
            <li>Erwartung In der Trefferliste enthalten alle Titel den
         Begriff Überleben</li>
        </ol>
    </div>
    <div category="yes" id="ts_1_1" scenario="ts_1_1.xml">
        <reuse system="iblk" category="yes"/>
        <!— Problem mit den unterschiedlichen Listen —>
        <reuse system="infoconnex" category="yes"/>
        <!— Solr hat keine Anzeige des Queries —>
        <reuse system="solr" category="no"/>
        <h2>Test 2(bs) Phrasensuche</h2>
        <p>Dieser Test überprüft die Phrasensuche</p>
        <ol>
            <li>Aktion In der Suche wird im Feld Titel nach ?geld und geltung?
         gesucht (doppelte
                Anführungstriche und Kleinschreibung beachten)</li>
            <li>Erwartung In der Anzeige Ihre Suche steht ?geld und geltung?</li>
            <li>Erwartung In der Ergebnisliste erscheinen 3 Treffer</li>
        </ol>
    </div>
    <div category="yes" id="ts_1_2" scenario="ts_1_2.xml">
        <!— Recent Years Filter ist nur für Sowiport vorhanden —>
        <reuse system="solr" category="can"/>
        <reuse system="infoconnex" category="can"/>
        <reuse system="iblk" category="can"/>
        <h2>Test 3(bs) Suche nach Soziologie im Zeitraum 1992-1997</h2>
        <p>Dieser Test überprüft die Suche in Zeiträumen und dient dem
         Abgleich der erzielten
            Treffermenge mit Erwartungswerten</p>
        <ol>
            <li>Aktion In der Suche wird die Checkbox Nur die letzten 4 Jahre
                          durchsuchen deaktiviert</li>
            <li>Aktion Im Feld Jahr wird die Zeichenfolge 1992-1997
                          eingetragen</li>
            <li>Aktion Im Feld Überall wird Soziologie eingetragen, die Suche
                          wird gestartet</li>
            <li>Erwartung Es werden mehr als 28000 Treffer gefunden</li>
            <li>Erwartung Es werden mehr als 4000 Treffer in Solis gefunden</li>
        </ol>
    </div>
 </out>
```

Test Set TS_2

Listing A.2: Excerpt from TS_2 : Tests for an Information Retrieval evaluation

```
<?oxygen RNGSchema="testsets.rnc" type="compact"?>
<out system="fast">
    <div category="yes" id="ts_2_0" scenario="ts_2_0.xml">
        <h2>Anzahl Dokumentzahlen</h2>
        <p>Einsatz CK > Nichteinsatz > 25 Anfragen: mehr Dokumente
                        grundsätzlich gefunden? </p>
    </div>
    <div category="yes" id="ts_2_1" scenario="ts_2_1.xml">
        <h2>Top10 / Top 20 A?nderun</h2>
        <p> Ändern sich bei Einsatz der CK die Dokumente in den Top 10 oder
                        Top 20 gerankten? </p>
    </div>
    <div category="yes" id="ts_2_2" scenario="ts_2_2.xml">
        <h2>Top 10 / Top 20 Reihenfolge</h2>
        <p> Ändert sich bei Einsatz der CK die Reihenfolge von gerankten
                        Dokumenten in den Top 10 /
            Top 20? </p>
    </div>
    <div category="yes" id="ts_2_3" scenario="ts_2_3.xml">
        <h2>Anzahl Datenbanken: insgesamt </h2>
        <p> Werden bei Einsatz der CK mehr Datenbanken in der Suche
                        angesprochen, d.h. werden
            Dokumente aus mehr Datenbanken gefunden? </p>
    </div>
 </out>
```

Test Set TS_{3a}

Listing A.3: Excerpt from TS_{3a} : Tests for s_1, defined by domain experts

```
<?oxygen RNGSchema="testsets.rnc" type="compact"?>
<out system="sowiport">
    <div category="yes" id="ts_3a_0" scenario="ts_3a_0.xml">
        <h2>Test 17 (ze) Test der Funktionalität des Suchfelds Jahr /
                        einfache Suche</h2>
        <p>Dieser Test prüft die Funktionalität des Suchfelds Jahr in der
                        einfachen Suche.</p>
        <ol>
            <li>Aktion: in der einfachen Suche Suche mit Titelstichwort:
                                Umweltschutz; Suche starten </li>
            <li>Erwartung: Treffermenge (T1) </li>
            <li>Aktion: in der einfachen Suche Suche mit Titelstichwort:
                                Umweltschutz</li>
            <li>Aktion: nur die letzten vier Jahre durchsuchen deaktivieren </li>
            <li>Aktion: Feld Jahr: 2004 2005 2006 2007 </li>
            <li>Erwartung: Treffermenge (T2), dabei T2= T1 </li>
            <li>Aktion: in der einfachen Suche Suche mit Titelstichwort:
                                Umweltschutz</li>
            <li>Aktion: nur die letzten vier Jahre durchsuchen deaktivieren</li>
            <li>Aktion: Feld Jahr: 2004-2007 </li>
            <li>Erwartung: Treffermenge (T3), dabei T3=T1=T2</li>
        </ol>
    </div>
    <div category="yes" id="ts_3a_1" scenario="ts_3a_1.xml">
        <h2>Test 18 (ze) Test Trunkierung Titelfeld</h2>
        <p>Dieser Test prüft die Trunkierung im Suchfeld Titel in der
                        einfachen Suche.</p>
        <ol>
            <li>Aktion: einfache Suche im Titelfeld: Drogen </li>
            <li>Erwartung: Treffermenge (T1) </li>
            <li>Aktion: einfache Suche im Titelfeld: Droge*</li>
            <li>Erwartung: Treffermenge (T2), T2&gt;T1</li>
        </ol>
    </div>
    <div category="yes" id="ts_3a_2" scenario="ts_3a_2.xml">
        <h2>Test 19a (ze) Test Verundung von Suchfeldern in der erweiterten
                        Suche / Literatur</h2>
        <p>Dieser Test prüft die Kombination von Suchfeldern
                        (Person und Überall) in der erweiterten
            Suche für den Informationstyp Literatur.</p>
        <ol>
```

```
            <li>Aktion: in der erweiterten Suche wird alle deaktiviert</li>
            <li>Aktion: in der erweiterten Suche wird Literatur ausgewählt</li>
            <li>Aktion: im Suchfeld Personen wird Mauss, Marcel eingegeben. </li>
            <li>Erwartung: Anzeige Ihre Suche zeigt Person: Mauss und Marcel;
                           Treffermenge (T1)</li>
            <li>Aktion: im Suchfeld Personen wird Mauss, Marcel eingegeben;
                           im Suchfeld Überall
                Soziologie</li>
            <li>Erwartung: Anzeige Ihre Suche zeigt Person: Mauss und Marcel
                           und Soziologie überall; Treffermenge (T2), dabei T2&lt;t1/></li>
        </ol>
    </div>
</out>
```

Test Set TS_{3b}

Listing A.4: Excerpt from TS_{3b} : Tests for s_3, defined by domain experts

```
<?oxygen RNGSchema="testsets.rnc" type="compact"?>
<out system="iblk">
    <div category="yes" id="ts_3b_0" scenario="ts_generic_search.xml">
        <h2>Test 1a(gl1) Suche mit Umlaut</h2>
        <p>Dieser Test überprüft die Auflösung des Umlauts ?ü? zu ?ue?</p>
        <ol>
            <li>Aktion In der erweiterten Suche in allen Datenbanken wird im
                           Feld Titel nach Überbevölkerung gesucht</li>
            <li>Erwartung Treffermenge T1</li>
            <li>Aktion In der erweiterten Suche in allen Datenbanken wird
                           im Feld Titel nach Ueberbevölkerung gesucht</li>
            <li>Erwartung Treffermenge T2, dabei T1=T2</li>
        </ol>
    </div>
    <div category="yes" id="ts_3b_1" scenario="ts_generic_search.xml">
        <h2>Test 1b(gl1) Suche mit Diakritika</h2>
        <p>Dieser Test überprüft die Suche mit dem Akzent ´</p>
        <ol>
            <li>Aktion In der erweiterten Suche in allen Datenbanken wird
                           im Feld Autor nach Méchet gesucht</li>
            <li>Erwartung Treffermenge T1 (notieren)</li>
            <li>Aktion In der erweiterten Suche in allen Datenbanken wird
                           im Feld Autor nach Mechet gesucht</li>
            <li>Erwartung Treffermenge T2, dabei T1=T2</li>
        </ol>
    </div>
    <div category="yes" id="ts_3b_2" scenario="ts_generic_search.xml">
        <h2>Test 1c(gl1) Suche mit Sonderzeichen</h2>
        <p>Dieser Test überprüft die Auflösung des ?ß? zu ?ss?</p>
        <ol>
            <li>Aktion In der erweiterten Suche in allen Datenbanken wird im
                           Feld Suche überall nach Aussenpolitik gesucht</li>
            <li>Erwartung Treffermenge T1 (notieren)</li>
            <li>Aktion In der erweiterten Suche in allen Datenbanken wird
                           im Feld Suche überall nach Außenpolitik gesucht</li>
            <li>Erwartung Treffermenge T2, dabei T1=T2</li>
        </ol>
    </div>
</out>
```

Test Set TS_{4a}

Listing A.5: Excerpt from TS_{4a} : Non-retrievals tests for system s_2

```
<?oxygen RNGSchema="testsets.rnc" type="compact"?>
<out system="infoconnex">
    <div category="external" id="ts_4a_0">
        <h2>Test der Browserfunktionalität</h2>
        <p>Läuft infoconnex auf allen Browsern und Benutzungsoberflächen?</p>
        <ol>
            <li>Internet Explorer</li>
```

```
            <li>Opera</li>
            <li>Mozilla Firefox</li>
            <li>Netscape</li>
        </ol>
    </div>


    <div category="external" id="ts_4a_2">
        <h2>Test der Darstellung</h2>
        <p>Stimmen die Style-Sheets? Bsp. Hinweisseite für Nutzer, die nicht
        über eine lizenzierte Bibliothek auf infoconnex zugreifen. "Es␣kann
␣␣␣␣␣␣␣nur␣eine␣Kurzinformation␣kostenlos␣angezeigt␣werden" (siehe
        Ausdruck)</p>
    </div>

    <div category="external" id="ts_4a_3">
        <h2>Test der Verfügbarkeitsrecherche</h2>
        <p>Ist die Verfügbarkeitsangabe richtig?</p>
        <ol>
            <li>Funktioniert der Link der Verfügbarkeitsinformation?</li>
            <li>Kann der Artikel wirklich bestellt werden?</li>
            <li>Kann man bei grüner Ampel den Artikel wirklich sofort
          herunterladen?</li>
            <li>Sind kostenlose Volltexte sofort verfügbar?</li>
            <li>Sind diese Volltexte lesbar (PDFs, Schrift,?)?</li>
        </ol>
    </div>

 </out>
```

Test Set TS_{4b}

Listing A.6: Excerpt from TS_{4b} : Retrieval tests for system s_2

```
<?oxygen RNGSchema="testsets.rnc" type="compact"?>
<out system="infoconnex">
    <div category="yes" id="ts_4b_3" scenario="ts_generic_search.xml">
        <h2>Anfrage 3</h2>
        <p>Jahr=1983:</p>
    </div>
    <div category="yes" id="ts_4b_59" scenario="ts_generic_search.xml">
        <h2>Anfrage 59</h2>
        <p>Überall="Soziale␣Ungleichheit"</p>
    </div>
    <div category="yes" id="ts_4b_136" scenario="ts_generic_filtered_search.xml">
        <h2>Anfrage 136</h2>
        <p>Filter: Datenquelle=SOLIS, FORIS; Sprache=englisch Suche:
        Titel=deprivation und Jahr=1993-2003</p>
    </div>
</out>
```

Test Set TS_5

Listing A.7: Excerpt from TS_5 : Tests derived from documented issues for s_1

```
<?oxygen RNGSchema="testsets.rnc" type="compact"?>
<out system="sowiport">
    <div category="external" id="ts_5_4" scenario="">
        <h2>Eintrag 4</h2>
        <p>Artikelübersichtsseite FQS unter Mac/Safari nicht schön (CSS)</p>
    </div>
    <div category="yes" id="ts_5_7" scenario="NOTMODELLED">
        <h2>Eintrag 7</h2>
        <p>in Suchanfragenanzeige Datenbank an sich prüfen kein endloses
                    Hinzufügen gleicher Informationstyp-Navigatoren</p>
    </div>
    <div category="yes" id="ts_5_13" scenario="NOTMODELLED">
        <h2>Eintrag 13</h2>
        <p>Zustandsanzeige: Fehlende Klammerung bei Suche: Schlagwort =
```

```
        (jugend or gewalt) and (ehe or familie) ergibt: Schlagwörter: Dasselbe
        Problem tritt auch bei Verwendung mehrerer Zeilen mit demselben
        Suchfeld auf.</p>
    </div>
    <div category="external" id="ts_5_14" scenario="">
        <h2>Eintrag 14</h2>
        <p>Schnellstartleiste: non-js-Version</p>
    </div>
</out>
```

Bibliography

[AB04] Johan Andersson and Geoff Bache. The Video Store Revisited Yet Again: Adventures in GUI Acceptance Testing. In Jutta Eckstein and Hubert Baumeister, editors, *Extreme Programming and Agile Processes in Software Engineering 5th International Conference XP 2004 Garmisch-Partenkirchen, Germany, June 2004*, volume 3092 of *Lecture Notes in Computer Science*. Springer, 2004.

[ABS00] Serge Abiteboul, Peter Buneman, and Dan Suciu. *Data on the Web: From Relations to Semistructured Data and XML*. Morgan Kaufman, 2000.

[AC04] Michal Antkiewicz and Krzysztof Czarnecki. FeaturePlugin: Feature Modeling Plug-in for Eclipse. In *OOPSLA'04 Eclipse Technology eXchange (ETX) Workshop,*, 2004.

[ALP05] Marcus Alanen, Torbjoern Lundkvist, and Ivan Porres. Comparison of Modeling Frameworks for Software Engineering. *Nordic Journal of Computing*, 12(4):321–342, 2005.

[And04a] Jennitta Andrea. Generative Acceptance Testing for Difficult-to-Test Software. In Jutta Eckstein and Hubert Baumeister, editors, *Extreme Programming and Agile Processes in Software Engineering*, volume 3092 of *Lecture Notes in Computer Science*, pages 29–37. Springer, 2004.

[And04b] Jennitta Andrea. Putting a Motor on the Canoo WebTest Acceptance Testing Framework. In Jutta Eckstein and Hubert Baumeister, editors, *Extreme Programming and Agile Processes in Software Engineering 5th International Conference XP 2004 Garmisch-Partenkirchen, Germany, June 2004*, volume 3092 of *Lecture Notes in Computer Science*. Springer, 2004.

[AOA05] Anneliese Andrews, Jeff Offutt, and Roger Alexander. Testing Web Applications by Modeling with FSMs. *Software Systems and Modeling*, 4(3):326–345, 2005.

[AOS04] O. Armbrust, M. A. Ochs, and B. Snoek. Stand der Praxis von Software-Tests und deren Automatisierung : Interviews und Online-Umfrage. Report; electronic publication, Fraunhofer Institut Experimentelles Software Engineering, 2004.

[Ara89] Guillermo Arango. Domain Analysis: From Art Form to Engineering Discipline. In *IWSSD '89: Proceedings of the 5th international workshop on Software specification and design*, pages 152–159, New York, NY, USA, 1989. ACM Press.

[Ara94] Guillermo Arango. A Brief Introduction to Domain Analysis. In *SAC '94: Proceedings of the 1994 ACM symposium on Applied computing*, pages 42–46, New York, NY, USA, 1994. ACM Press.

[BA04] Kent Beck and Cynthia Andres. *Extreme Programming Explained: Embrace Change (2nd Edition)*. Addison-Wesley Professional, 2004.

[Bas96] Victor R. Basili. The role of experimentation in software engineering: past, current, and future. In *ICSE '96: Proceedings of the 18th international conference on Software engineering*, pages 442–449, Washington, DC, USA, 1996. IEEE Computer Society.

[BBN02] Mark Blackburn, Robert Busser, and Aaron Nauman. Understanding the Generations of Test Automation, 2002.

[BBN04] M. Blackburn, R. Busser, and A. Nauman. Why Model-Based Test Automation is Different and What You Should Know to Get Started. In *International Conference on Practical Software Quality*, 2004.

[BC87] Nicholas J. Belkin and W. Bruce Croft. *Retrieval Techniques*, pages 109–145. Elsevier Science Inc., New York, NY, USA, 1987.

[BC90] Gilad Bracha and William Cook. Mixin-based Inheritance. In *OOPSLA/ECOOP '90: Proceedings of the European conference on object-oriented programming on Object-oriented programming systems, languages, and applications*, pages 303–311, New York, NY, USA, 1990. ACM.

[BC05] Felix Bachmann and Paul C. Clements. Variability in Software Product Lines. Technical report, Carnegie Mellon Software Engineering Institute, 2005.

[BCR94] Victor R. Basili, Gianluigi Caldiera, and H. Dieter Rombach. *Encyclopedia of Software Engineering*, chapter The Goal Question Metric Approach, pages 528–532. John Wiley & Sons, 1994.

[BDG+03] Paul Baker, Zhen Ru Dai, Jens Grabowski, Øystein Haugen, Serge Lucio, Eric Samuelsson, and Ina Schieferdecker Clay E. Williams. The UML 2.0 Testing Profile, 2003.

[Bec02] Kent Beck. *Test Driven Development: By Example*. Addison-Wesley, 2002.

[Bec08] Steffen Becker. *Coupled Model Transformations for QoS Enabled Component-Based Software Design. PhD thesis, Carl von Ossietzky Universität Oldenburg, Oldenburg, Germany, 2008.*

[Bei90] Boris Beizer. Software testing techniques (2nd ed.). *Van Nostrand Reinhold Co., New York, NY, USA, 1990.*

[Bei95] Boris Beizer. Black-Box Testing: Techniques for Functional Testing of Software and Systems. *John Wiley & Sons, Inc., New York, NY, USA, 1995.*

[Ber07] Antonia Bertolino. Software Testing Research: Achievements, Challenges, Dreams. In 29th Int. Conference on Software Engineering - Future of Software Engineering(FOSE'07)*, 2007.*

[Béz05] *Jean Bézivin. On the Unification Power of Models.* Software and Systems Modeling, *4(2):171–188, 2005.*

[BFFR06] *Benoit Baudry, Franck Fleurey, Robert France, and Raghu Reddy. Exploring the Relationship between Model Composition and Model Transformation. In* Report of the 7th International Workshop on Aspect-Oriented Modeling, *2006.*

[BFK+99] *Joachim Bayer, Oliver Flege, Peter Knauber, Roland Laqua, Dirk Muthig, Klaus Schmid, Tanya Widen, and Jean-Marc DeBaud. PuLSE: a methodology to develop software product-lines. In* SSR '99: Proceedings of the 1999 symposium on Software reusability, *pages 122–131, New York, NY, USA, 1999. ACM Press.*

[BG03a] *Antonia Bertolino and Stefania Gnesi. PLUTO: A Test Methodology for Product Families. In Frank van der Linden, editor,* Proceedings of the Fifth International Workshop on Product Family Engineering (PFE-5), *LNCS 3014, Siena, Italy, 2003. Springer Verlag.*

[BG03b] *Antonia Bertolino and Stefania Gnesi. Use case-based testing of product lines. In* ESEC/FSE-11: Proceedings of the 9th European software engineering conference held jointly with 11th ACM SIGSOFT international symposium on Foundations of software engineering, *pages 355–358, New York, NY, USA, 2003. ACM Press.*

[BGH07] *C. Titus Brown, Grig Gheorghiu, and Jason R. Huggins.* An Introduction to Testing Web Applications with twill and Selenium. *O'Reilly Media, Inc., 2007.*

[Bin99] *Robert V. Binder.* Testing Object-Oriented Systems: Models, Patterns, and Tools. *Addison-Wesley, 1999.*

[Bla80] *D.C. Blair. Searching biases in large interactive document retrieval systems.* Journal of the American Society for Information Science, *31(4):271–277, 1980.*

[BLC05] *L. C. Briand, Y. Labiche, and J. Cui. Automated support for deriving test requirements from UML statecharts.* Software and Systems Modeling, *4:399–423, 2005.*

[BMM04] *A. Bertolino, E. Marchetti, and H. Muccini. Introducing a Reasonably Complete and Coherent Approach for Model-based Testing, 2004.*

[BMR+96] *Frank Buschmann, Regine Meunier, Hans Rohnert, Peter Sommerlad, and Michael Stal.* Pattern-Oriented Software Architecture Volume 1: A System of Patterns. *Wiley, 1996.*

[BPR03] *Jewgenij Botaschanjan, Markus Pister, and Bernhard Rumpe. Testing Agile Requirements Models. In* In: Proceedings of the First Hangzhou-Lübeck Conference on Software Engineering (HL-SE 03), *2003.*

[BR05] *M. Broy and A. Rausch. Das neue V-Modell® XT.* Informatik-Spektrum, *28(3):220–229, 2005.*

[Bro87] *Fred P. Brooks. No Silver Bullet - essence and accident in software Engineering Computer.* Computer, *20(4):10–19, 1987.*

[Bro04] *Alan W. Brown. Model driven architecture: Principles and practice.* Software and System Modeling, *3(4):314–327, 2004.*

[BSL99] *Victor R. Basili, Forrest Shull, and Filippo Lanubile. Building Knowledge through Families of Experiments.* IEEE Trans. Softw. Eng., *25(4):456–473, 1999.*

[BSM+03] *Frank Budinsky, David Steinberg, Ed Merks, Raymond Ellersick, and Timothy J. Grose.* Eclipse Modeling Framework. *Addison Wesley Professional, 2003.*

[BY01] *Luciano Baresi and Michal Young. Test Oracles. Technical report, Department of Computer and Information Science, University of Oregon, 2001.*

[BYRN99] *Ricardo A. Baeza-Yates and Berthier Ribeiro-Neto.* Modern Information Retrieval. *Addison-Wesley Longman Publishing Co., Inc., Boston, MA, USA, 1999.*

[CA05] *Krzysztof Czarnecki and Michal Antkiewicz. Mapping Features to Models: A Template Approach Based on Superimposed Variants. In* International Conference on Generative Programming and Component Engineering, *2005.*

[CAJ01] *R.B. Chase, N.J. Aquilano, and F.R. Jacobs.* Operations Management for Competitive Advantage. *McGraw-Hill Irwin, 2001.*

[CAK06] *Krzysztof Czarnecki, Michal Antkiewicz, and Chang Hwan Peter Kim. Multilevel customization in application engineering.* Commun. ACM, *49(12):60–65, 2006.*

[CCD04] *Alessandra Cavarra, Charles Crichton, and Jim Davies. A Method for the Automatic Generation of Test Suites from Object Models.* Information & Software Technology, *46(5):309–314, 2004.*

[CE00] *Krysztof Czarnecki and Ulrich W. Eisenecker.* Generative Programming. Methods, Tools and Applications. *Addison-Wesley, 2000.*

[CH03] *Krzysztof Czarnecki and Simon Helsen. Classification of Model Transformation Approaches. In* OOPSLA'03 Workshop on Generative Techniques in the Context of Model-Driven Architecture, *2003.*

[Cha03] *S. Chakrabarti.* Mining the Web: Discovering Knowledge from Hypertext Data. *Morgan Kaufmann, 2003.*

[Che05a] *T. Y. Chen. Are Successful Test Cases Useless or Not? In Ralf Reussner, Johannes Mayer, Judith A. Stafford, Sven Overhage, Steffen Becker, and Patrick J. Schroeder, editors,* QoSA/SOQUA, *volume 3712 of* Lecture Notes in Computer Science, *pages 2–3. Springer, 2005.*

[CHE05b] *Krzysztof Czarnecki, Simon Helsen, , and Ulrich Eisenecker. Staged Configuration Through Specialization and Multi-Level Configuration of Feature Models.* Software Process Improvement and Practice, special issue on "Software Variability: Process and Management,, *143 - 169:10(2), 2005.*

[CHE05c] *Krzysztof Czarnecki, Simon Helsen, and Ulrich Eisenecker. Formalizing Cardinality-based Feature Models and their Specialization.* Software Process Improvement and Practice, special issue of best papers from SPLC04, *10(1):7 – 29, 2005.*

[CHW98] *James Coplien, Daniel Hoffman, and David Weiss. Commonality and Variability in Software Engineering.* IEEE Software, *15(6):37–45, 1998.*

[CKK06] *Krzysztof Czarnecki, Chang Hwan Peter Kim, and Karl Trygve Kalleberg. Feature Models Are Views on Ontologies. In* Proceedings of 10th International Software Product Line Conference (SPLC 2006), *pages 41–51, 2006.*

[CLW06] *James R. Cordy, Ralf Lämmel, and Andreas Winter. Executive Summary – Transformation Techniques in Software Engineering. In* Transformation Techniques in Software Engineering, *volume 05161 of* Dagstuhl Seminar Proceedings. *Internationales Begegnungs- und Forschungszentrum (IBFI), Schloss Dagstuhl, Germany, Dagstuhl, 2006.*

[CN01] *Paul Clements and Linda M. Northrop.* Software Product Lines : Practices and Patterns. *Addison Wesley, 3rd edition, 2001.*

[Con04] *Chris Condron. Domain Approach to Test Automation of Product Lines. In Birgit Geppert, Charles Krueger, and Jenny Li, editors,* Proceedings of the International Workshop on Software Product Line Testing (SPLiT 2004), *pages 27–35, Boston, MA, 2004.*

[Coo73] *W.S. Cooper. On Selecting a Measure of Retrieval Effectiveness.* Journal of the American Society for Information Science, *24(6):413–424, 1973.*

[CPRS04] *V. Cechticky, A. Pasetti, O. Rohlik, and W. Schaufelberger. XML-Based Feature Modelling. In* Software Reuse: Methods, Techniques and Tools: 8th International Conference, ICSR, *pages 5–9. Springer, 2004.*

[Cro79] *P.B. Crosby.* Quality is Free: The Art of Making Quality Certain. *McGraw-Hill, 1979.*

[CT06] *Jordi Cabot and Ernest Teniente. Constraint Support in MDA tools: a Survey. In Arend Rensink and Jos Warmer, editors,* Model Driven Architecture - Foundations and Applications Second European Conference, ECMDA-FA 2006 Bilbao, Spain, July 10-13, 2006, *number 4066 in Lecture Notes in Computer Science. Springer, 2006.*

[Cza05] *Krzysztof Czarnecki. Overview of Generative Software Development. In J.-P. Banâtre et al:, editor,* Unconventional Programming Paradigms (UPP) 2004, *volume 3566 of* Lecture Notes of Computer Science, *pages 313–328, Mont Saint-Michel, France, 2005. Springer.*

[Dai04] *Zhen Ru Dai. Model-Driven Testing with UML 2.0. In* Second European Workshop on Model Driven Architecture (MDA) EWMDA, *pages 179–188, 2004.*

[DER05] *Hyunsook Do, Sebastian Elbaum, and Gregg Rothermel. Supporting Controlled Experimentation with Testing Techniques: An Infrastructure and its Potential Impact.* Empirical Software Engineering, *10(4):405–435, 2005.*

[DFG06] *Vincenzo Deufemia, Filomena Ferrucci, and Carmine Gravino. Constructing Meta-CASE Workbenches by Exploiting Visual Language Generators.* IEEE Trans. Softw. Eng., *32(3):156–175, 2006. Member-Gennaro Costagliola.*

[Dij69] *E.W. Dijkstra.* Notes on structured programming. *Technological University Eindhoven, Netherlands Department of Mathematics, [Eindhoven], 1969.*

[DMBK01] *Arie Deursen, Leon Moonen, Alex Bergh, and Gerard Kok. Refactoring Test Code, 2001.*

[DMG07] *Paul Duvall, Steve Matyas, and Andrew Glover.* Continuous Integration: Improving Software Quality and Reducing Risk. *Addison-Wesley Professional, 2007.*

[Dmi04] *Sergey Dmitriev. Language Oriented Programming: The Next Programming Paradigm, 2004.*

[DS99] *Jean-Marc DeBaud and Klaus Schmid. A systematic approach to derive the scope of software product-lines. In* ICSE '99: Proceedings of the 21st international conference on Software engineering, *pages 34–43, Los Alamitos, CA, USA, 1999. IEEE Computer Society Press.*

[EBI06] *Michael Ellims, James Bridges, and Darrel C. Ince. The Economics of Unit Testing.* Empirical Softw. Engg., *11(1):5–31, 2006.*

[Edw68] *C.D. Edwards. The meaning of quality.* Quality Progress, *1(10):36–39, 1968.*

[EFW01] *Ibrahim K. El-Far and James A. Whittaker. Model-Based Software Testing. In J.J. Marciniak, editor,* Encyclopedia on Software Engineering, *pages 825–837. Wiley, 2001.*

[Eib00] *M. Eibl.* Visualisierung im Document Retrieval: theoretische und praktische Zusammenführung von Softwareergonomie und Graphik Design. *PhD thesis, GESIS, Informationszentrum Sozialwissenschaften, 2000.*

[ES06] *Matthew Emerson and Janos Sztipanovits. Techniques for Metamodel Composition. In* Proceedings of 6th OOPSLA Workshop on Domain-Specific Modeling (DSM'06), *number 37 in Computer Science and Information System Reports, Technical Reports. University of Jyväskylä, 2006.*

[ES07] *Robert B. Evans and Alberto Savoia. Differential testing: a new approach to change detection. In* ESEC-FSE '07: Proceedings of the the 6th joint meeting of the European software engineering conference and the ACM SIGSOFT symposium on The foundations of software engineering, *pages 549–552, New York, NY, USA, 2007. ACM Press.*

[Eva04] *Eric Evans.* Domain-Driven Design: Tackling Complexity in the Heart of Software. *Addison-Wesley, 2004.*

[FER08] *F. Freiling, I. Eusgeld, and R. Reussner, editors.* Dependability Metrics. *Lecture Notes in Computer Science. Springer-Verlag, 2008.*

[Fow03] *Martin Fowler.* Patterns of Enterprise Application Architecture. *Addison-Wesley, 2003.*

[Fow05a] *Martin Fowler. A Language Workbench in Action - MPS., 2005.*

[Fow05b] *Martin Fowler. Language Workbenches and Model Driven Architecture, 2005.*

[Fow05c] *Martin Fowler. Language Workbenches: The Killer-App for Domain Specific Languages?, 2005.*

[Fow06] *Martin Fowler. Specification By Example, 2006.*

[FR07] *Robert France and Bernhard Rumpe. Model-Driven Development of Complex Software: A Research Roadmap. In* Future of Software Engineering(FOSE'07), *2007.*

[fS01] *International Organization for Standardization. ISO 9126 Software engineering – Product quality – Part 1: Quality model, 2001.*

[Gäl06] *Markus Gälli.* Modeling Examples to Test and Understand Software. *PhD thesis, Universität Bern, 2006.*

[GBS01] *Jilles Van Gurp, Jan Bosch, and Mikael Svahnberg. On the Notion of Variability in Software Product Lines. In* WICSA '01: Proceedings of the Working IEEE/IFIP Conference on Software Architecture (WICSA'01), *page 45, Washington, DC, USA, 2001. IEEE Computer Society.*

[GEMT06] *Javier J. Gutierrez, Maria J. Escalona, Manuel Mejias, and Jesus Torres. Generation of test cases from functional requirements. A survey. Technical report, Department of Computer Languages and Systems, University of Seville, 2006.*

[GF00] *Dorothy Graham and Mark Fewster, editors.* Software Test Automation: Effective Use of Test Execution Tools. *Addison-Wesley, 2000.*

[GHJV95] *Erich Gamma, Richard Helm, Ralph Johnson, and John Vlissides.* Design Patterns: Elements of Reusable Object-Oriented Softwar. *Addison-Wesley Professional, 1995.*

[GKR+06] *H. Grönniger, H. Krahn, B. Rumpe, M. Schindler, and S. Völkel. MontiCore 1.0 - Ein Framework zur Erstellung und Verarbeitung domänenspezifischer Sprachen. Technical report, Technische Universität Braunschweig, Carl-Friedrich-Gauss-Fakultät für Mathematik und Informatik, 2006.*

[GLRW04] *Birgit Geppert, Jenny Li, Frank Rößler, and David M. Weiss. Towards Generating Acceptance Test Cases for Product Lines. In* Proceedings of the Eighth International Conference on Software Reuse, *2004.*

[GS04] *Jack Greenfield and Keith Short.* Software Factories: Assembling Applications with Patterns, Models, Frameworks, and Tools. *Wiley, 2004.*

[Has97] *W. Hasselbring. Federated integration of replicated information within hospitals.* International Journal on Digital Libraries, *1(3):192–208, 1997.*

[Has02] *W. Hasselbring. Web Data Integration for E-Commerce Applications.* IEEE MULTIMEDIA, *pages 16–25, 2002.*

[Hei07] *Ehrhardt F. Heinold. Virtuelle Fachbibliotheken im System der überregionalen Literatur- und Informationsversorgung : Studie zu Angebot und Nutzung der Virtuellen Fachbibliotheken . Technical report, Heinold, Spiller & Partner Unternehmensberatung, 2007.*

[Her03] *Jack Herrington.* Code Generation in Action. *Manning Publications, July 2003.*

[HHK+05] *Jo E. Hannay, Ove Hansen, Vigdis By Kampenes, Amela Karahasanovic, Nils-Kristian Liborg, and Anette C. Rekdal. A Survey of Controlled Experiments in Software Engineering.* IEEE Trans. Softw. Eng., *31(9):733–753, 2005. Member-Dag I. K. Sjoberg.*

[HR04] *D. Harel and B. Rumpe. Meaningful Modeling: What's the Semantics of Semantics?* Computer, *37(10):64–72, October 2004.*

[HS00] *Daniel Hoffman and Paul Strooper. Prose + Test Cases = Specifications. In* TOOLS '00: Proceedings of the Technology of Object-Oriented Languages and Systems (TOOLS 34'00), *page 239, Washington, DC, USA, 2000. IEEE Computer Society.*

[HS07] *Sabine Heinz and Maximilian Stempfhuber. Eine informationsarchitektur für wissenschaftliche fachportale in vascoda. In Achim Oßwald, Maximilian Stempfhuber, and Christian Wolff, editors,* Open Innovation: neue Perspektiven im Kontext von Information und Wissen; Beiträge des 10. internationalen Symposiums für Informationswissenschaft und der 13. Jahrestagung der IuK-Initiative Wissenschaft, *Schriften zur Informationswissenschaft, pages 485–507, 2007.*

[HST07] *Patrick Heymans, Pierre-Yves Schobbens, and Jean-Christophe Trigau. Towards the Comparative Evaluation of Feature Diagram Languages. In Tomi Männistö, Eila Niemelä, and Mikko Raatikainen, editors,* Software and Service Variability Management Workshop - Concepts, Models, and Tools, *number 3 in Research Reports, pages 1–16. Helsinki University of Technology, Software Business and Engineering Institute, 2007.*

[Hud96] *Paul Hudak. Building domain-specific embedded languages.* ACM Computing Surverys, *28(4es):196, 1996.*

[HVR04] *Jean Hartmann, Marlon Vieira, and Axel Ruder. A UML-based Approach for Validating Product Lines. In Birgit Geppert, Charles Krueger, and Jenny Li, editors,* Proceedings of the International Workshop on Software Product Line Testing (SPLiT 2004), *pages 58–65, Boston, MA, 2004.*

[JG88] *J.M. Juran and F.M. Gryna, editors.* Juran's quality control handbook. *McGraw-Hill, 1988.*

[JS05] *David Janzen and Hossein Saiedian. Test-Driven Development: Concepts, Taxonomy, and Future Direction.* Computer, *38(9):43–50, 2005.*

[Kan02] *Stephen H. Kan.* Metrics and Models in Software Quality Engineering. *Addison-Wesley Longman, 2nd edition, 2002.*

[Kan03] *Cem Kaner. What is a good test case? In* Software Testing Analysis & Review Conference (STAR) East, *Orlando, FL, May 12-16 2003.*

[KBP01] *Cem Kaner, James Bach, and Bret Pettichord.* Lessons Learned in Software Testing. *Wiley, 2001.*

[KC05] *Chang Hwan Peter Kim and Krzysztof Czarnecki. Synchronizing Cardinality-Based Feature Models and their Specializations. In A. Hartman and D. Kreische, editors,* Proceedings of European Conference on Model Driven Architecture – Foundations and Applications (ECMDA-FA'05), *number 3748 in Lecture Notes in Computer Science, pages 331–348, Nuremberg, Germany, 2005. Springer.*

[KCH+90] *K. Kang, S. Cohen, J. Hess, W. Novak, and A. Peterson. Feature-oriented domain analysis (FODA) feasibility study. Technical Report CMU/SEI-90-TR-021, Software Engineering Institute, Carnegie-Mellon University,, November 1990.*

[KDJ04] *Barbara A. Kitchenham, Tore Dyba, and Magne Jorgensen. Evidence-Based Software Engineering. In* ICSE '04: Proceedings of the 26th International Conference on Software Engineering, *pages 273–281, Washington, DC, USA, 2004. IEEE Computer Society.*

[KG03] *Daryl Kulak and Eamonn Guiney.* Use Cases: Requirements in Context. *Addison-Wesley Longman Publishing Co., Inc., Boston, MA, USA, 2003.*

[KK02] *Dimitris Karagiannis and Harald Kühn. Metamodelling Platforms. In* Proceedings of the Third International Converence EC-Web, *2002.*

[KM06] *Ronny Kolb and Dirk Muthig. Making testing product lines more efficient by improving the testability of product line architectures. In* ROSATEA '06: Proceedings of the ISSTA 2006 workshop on Role of software architecture for testing and analysis, *pages 22–27, New York, NY, USA, 2006. ACM Press.*

[KMA03] *Stefan Kettemann, Dirk Muthig, and Michalis Anastasopoulos. Product Line Implementation Technologies : Component Technology View. Technical Report IESE-Report No. 015.03/E, Fraunhofer Institut Experimentelles Software Engineering, 2003.*

[Kol00] *M. Koll. Information Retrieval.* BULLETIN-AMERICAN SOCIETY FOR INFORMATION SCIENCE, *26(2):16–17, 2000.*

[Koz08] *Heiko Koziolek.* Parameter Dependencies for Reusable Performance Specifications of Software Components. *PhD thesis, Carl von Ossietzky Universität Oldenburg, January 2008.*

[Kra96] *J. Krause. Informationserschliessung und -bereitstellung zwischen Deregulation, Kommerzialisierung und weltweiter Vernetzung - Schalenmodell. Technical Report 6, GESIS-IZ Sozialwissenschaften, September 1996.*

[Kra06a] *Jürgen Krause. Shell Model, Semantic Web and Web Information Retrieval. In Ilse Harms, Heinz-Dirk Luckhardt, and Hans W. Giessen, editors,* Information und Sprache: Beiträge zu Informationswissenschaft, Computerlinguistik, Bibliothekswesen und verwandten Fächern; Festschrift für Harald H. Zimmermann., *pages 95 – 106. München: Saur, 2006.*

[Kra06b] *Jürgen Krause. Visual Interaction on the Basis of the WOB-Model. In Reinhad Rapp, Peter Seldmeier, and Gisela Zunker-Rapp, editors,* Perspectives on Cognition - A Festschrift for Manfred Wettler. *Pabst Science Publishers, 2006.*

[Kru03] *P. Kruchten.* The Rational Unified Process: An Introduction. *Addison-Wesley Professional, 2003.*

[KRV06] *Holger Krahn, Bernhard Rumpe, and Steven Völkel. Roles in Software Development using Domain Specific Modelling Languages. In* Proceedings of 6th OOPSLA Workshop on Domain-Specific Modeling (DSM'06), *number 37 in Computer Science and Information System Reports, Technical Reports. University of Jyväskylä, 2006.*

[KWD05] *Soon-Kyeong Kim, Luke Wildman, and Roger Duke. A UML Approach to the Generation of Test Sequences for Java-Based Concurrent Systems. In* Australian Software Engineering Conference, *pages 100–109. IEEE Computer Society, 2005.*

[Lar04] *Craig Larman.* Applying UML and Patterns: An Introduction to Object-Oriented Analysis and Design and the Unified Process. *Pretentice Hall PTR, 3rd edition, 2004.*

[LB03] *Craig Larman and Victor R. Basili. Iterative and Incremental Development: A Brief History.* Computer, *36(6):47–56, 2003.*

[Lew07] *Dirk Lewandowski. Mit welchen kennzahlen lässt sich die qualität von suchmaschinen messen? In Marcel Machill and Markus Beiler, editors,* Die Macht der Suchmaschinen / The Power of Search Engines. *Herbert von Halem Verlag, Köln, 2007.*

[LGR06] *Neil Loughran, Iris Groher, and Awais Rashid. Good Practice Guidelines for Code Generation in Software Product Line Engineering. In Paul Clements and Dirk Muthig, editors,* Variability Management – Working with Variability Mechanisms:Proceedings of the Workshop held in conjunction with the 10th Software Product Line Conference (SPLC-2006), *2006.*

[LH07] *Dirk Lewandowski and Nadine Hoechstoetter. Qualitätsmessung bei suchmaschinen: System- und nutzerbezogene evaluationsmaße.* Informatik Spektrum, *30(3), 2007.*

[LHRM07] *Ping Li, Toan Huynh, Marek Reformat, and James Miller. A practical approach to testing GUI systems.* Empirical Software Engineering, *12(4):331 – 357, August 2007.*

[LJX$^+$04] *Wang Linzhang, Yuan Jiesong, Yu Xiaofeng, Hu Jun, Li Xuandong, and Zheng Guoliang. Generating Test Cases from UML Activity Diagram based on Gray-Box Method.* apsec, *00:284–291, 2004.*

[Lud03] *Jochen Ludewig. Models in software engineering – an introduction.* Software and Systems Modeling, *2:5–14, 2003.*

[Man08] *Thomas Mandl. Automatische Bewertung der Qualität von Web-Seiten. Habilitationsschrift, 2008.*

[Mar07] *Anne Martens. Empirical Validation of the Model-Driven Performance Prediction Approach Palladio. Master's thesis, Carl von Ossietzky Unversität Oldenburg, November 2007.*

[Mat04] *Mari Matinlassi. Comparison of Software Product Line Architecture Design Methods: COPA, FAST, FORM, KobrA and QADA. In* ICSE '04: Proceedings of the 26th International Conference on Software Engineering, *pages 127–136, Washington, DC, USA, 2004. IEEE Computer Society.*

[May06] *Philipp Mayr. IZ-Arbeitsbericht Informationsangebote für das Wissenschaftsportal vascoda - eine Bestandsaufnahme. Technical Report 37, In formationszentrum Sozialwissenschaften der Arbeitsgemeinschaft Sozialwissenschaftlicher Institute e.V. (ASI), 2006.*

[MBN03] *Atif M. Memon, Ishan Banerjee, and Adithya Nagarajan. What Test Oracle Should I Use for Effective GUI Testing? In* 18th IEEE International Conference on Automated Software Engineering (ASE 2003), 6-10 October 2003, Montreal, Canada, *pages 164–173. IEEE Computer Society, 2003.*

[MC05a] *R. Mugridge and W. Cunningham. Agile Test Composition. In* Extreme Programming And Agile Processes in Software Engineering: 6th International Conference, XP 2005, Sheffield, UK, *volume 3556 of* Lecture Notes in Computer Science, *pages 137–144. Springer, June 18-23 2005.*

[MC05b] *Rick Mugridge and Ward Cunningham.* FIT for Developing Software. Framework for Integrated Tests. *Prentice Hall PTR, 2005.*

[McC04] *Steve McConnell.* Code Complete, Second Edition. *Microsoft Press, 2004.*

[McK98] *William M. McKeeman. Differential Testing for Software.* Digital Technical Journal, *10(1):100–107, 1998.*

[McN03] *Ashley McNeile. MDA: The Vision with the Hole?, 2003.*

[Mei06a] *Juergen Meister.* Produktgetriebene Entwicklung von Software-Produktlinien am Beispiel analytischer Anwendungssoftware. *PhD thesis, Carl von Ossietzky Universitat Oldenburg, 2006.*

[Mei06b] *Jürgen Meister. Software-Produktlinien. In Ralf Reussner and Wilhelm Hasselbring, editors,* Handbuch der Software-Architektur, *pages 370–393. dpunkt.verlag, 2006.*

[Mem01] *A.M. Memon.* A COMPREHENSIVE FRAMEWORK FOR TESTING GRAPHICAL USER INTERFACES. *PhD thesis, University of Pittsburgh, 2001.*

[Mes07] *Gerard Meszaros.* xUnit Test Patterns. *Addison Wesley, 2007.*

[MHS05] *Marjan Mernik, Jan Heering, and Anthony M. Sloane. When and how to develop domain-specific languages.* ACM Comput. Surv., *37(4):316–344, 2005.*

[MP08] *Philipp Mayr and Vivien Petras. Cross-concordances: terminology mapping and its effectiveness for Information Retrieval. In* IFLA World Library and Information Congress, *2008.*

[MPS01] *Atif M. Memon, Martha E. Pollack, and Mary Lou Soffa. Hierarchical GUI Test Case Generation Using Automated Planning.* IEEE Trans. Softw. Eng., *27(2):144–155, 2001.*

[MRS08] *Christopher D. Manning, Prabhakar Raghavan, and Hinrich Schütze.* Introduction to Information Retrieval. *Cambridge University Press., 2008.*

[MS05] *Varvana Myllännistö and Timo Soininen. Kumbang Configurator - A Configuration Tool for Software Product Families. Presented at the . In* International Joint Conference on Artificial Intelligence 2005 Configuration workshop,, *2005.*

[MSBT04] *Glenford J. Myers, Corey Sandler, Tom Badgett, and Todd M. Thomas.* The Art of Software Testing, Second Edition. *John Wiley & Sons, 2nd edition, 2004.*

[MSM04] *John D. McGregor, Prakash Sodhani, and Sai Madhavapeddi. Testing Variability in a Software Product Line. In* Proceedings of the International Workshop on Software Product Line Testing (SPLiT 2004), *2004.*

[MSMM00] *Jonathan I. Maletic, Khalid S. Soliman, Manuel A. Moreno, and William M. Mercer. Identification of Test Cases from Business Requirements of Software Systems, 2000.*

[MSU04] *Stephen J. Mellor, Kendall Scott, and Axel Uhl.* MDA Distilled: Principles of Model-Driven Architecture. *Addison-Wesley, 2004.*

[MSW05] *Philipp Mayr, Maximilian Stempfhuber, and Anne-Kathrin Walter. Auf dem Weg zum wissenschaftlichen Fachportal – Modellbildung und Integration heterogener Informationssammlungen. In* 27. DGI-Online-Tagung in Frankfurt am Main, *2005.*

[Mug04] *Rick Mugridge. Test Driving Custom Fit Fixtures. In Jutta Eckstein and Hubert Baumeister, editors,* Extreme Programming and Agile Processes in Software Engineering 5th International Conference XP 2004 Garmisch-Partenkirchen, Germany, June 2004, *volume 3092 of* Lecture Notes in Computer Science. *Springer, 2004.*

[Mut03] *P. Mutschke. Mining Networks and Central Entities in Digital Libraries. A Graph Theoretic Approach Applied to Co-author Networks.* LECTURE NOTES IN COMPUTER SCIENCE, *pages 155–166, 2003.*

[Myl05] *John Mylopoulos. Goal-Oriented Requirements Engineering.* apsec, *0:3, 2005.*

[Nei89] *James M. Neighbors. Draco: A Method for Engineering Reusable Software Systems. In Ted J. Biggerstaff and Alan J. Perlis, editors,* Software Reusability – Concepts and Models, *volume I, pages 295–319. ACM Press, 1989.*

[NF06] *Clementine Nebut and Franck Fleurey. Automatic Test Generation: A Use Case Driven Approach.* IEEE Trans. Softw. Eng., *32(3):140–155, 2006. Member-Yves Le Traon and Member-Jean-Marc Jezequel.*

[Nie94] *J. Nielsen. Guerrilla HCI: Using Discount Usability Engineering to Penetrate the Intimidation Barrier. In Randolph G. Bias and Deborah J. Mayhew, editors,* Cost-Justifying Usability, *pages 245–272. Morgan Kaufmann, 1994.*

[Nie01] *J. Nielsen. How to Conduct a Heuristic Evaluation. 2001.*

[Nor06] *Donald Norman. Why doing user observations first is wrong.* interactions, *13(4):50–ff, 2006.*

[Par76] *David L. Parnas. On the Design and Development of Program Families.* IEEE Transactions on Software Engineering, *2(1):1 – 9, March 1976.*

[PBS07] *Vivien Petras, Stefan Baerisch, and Maximillian Stempfhuber. The Domain-Specific Track at CLEF 2007. In Alessandro Nardi and Carol Peters, editors,* Working Notes for the CLEF 2007 Workshop, 19-21 September, Budapest, Hungary, *2007.*

[PCFJ03] *Teade Punter, Marcus Ciolkowski, Bernd Freimut, and Isabel John. Conducting On-line Surveys in Software Engineering. In* ISESE '03: Proceedings of the 2003 International Symposium on Empirical Software Engineering, *page 80, Washington, DC, USA, 2003. IEEE Computer Society.*

[PCSF04] *C. Michael Pilato, Ben Collins-Sussman, and Brian W. Fitzpatrick.* Version Control with Subversion. *O'Reilly Media, 2004.*

[Pet06] *Vivien Petras.* Translating Dialects in Search: Mapping between Specialized Languages of Discourse and Documentary Languages. *PhD thesis, University of California, Berkeley, 2006.*

[PH04] *S. Pedersen and W. Hasselbring. Interoperabilität für Informationssysteme im Gesundheitswesen auf Basis medizinischer Standards.* Informatik-Forschung und Entwicklung, *18(3):174–188, 2004.*

[PHM06] *Herbert Prähofer, Dominik Hurnaus, and Hanspeter Mössenböck. Building End-User Programming Systems Based on a Domain-Specific Language. In* Proceedings of 6th OOPSLA Workshop on Domain-Specific Modeling (DSM'06)*, number 37 in Computer Science and Information System Reports, Technical Reports. University of Jyväskylä, 2006.*

[PM02] *Thomas Patzke and Dirk Muthig. Product Line Implementation Technologies : Programming Language View. Technical Report IESE-Report No. 057.02/E, Fraunhofer Institut Experimentelles Software Engineering, 2002.*

[PM03] *Thomas Patzke and Dirk Muthig. Product Line Implementation with Frame Technology: A Case Study. Technical Report IESE-Report No. 018.03/E, Fraunhofer Institut Experimentelles Software Engineering, 2003.*

[PM06] *Klaus Pohl and Andreas Metzger. Software product-line testing.* Communication of the ACM, *49(12):78–81, 2006.*

[PNT06] *Andreas Prinz, Jan P. Nytun, and Merete S. Tveit. Automatic Generation of Modelling Tools. In Arend Rensink and Jos Warmer, editors,* Model Driven Architecture - Foundations and Applications Second European Conference, ECMDA-FA 2006 Bilbao, Spain, July 10-13, 2006*, number 4066 in Lecture Notes in Computer Science. Springer, 2006.*

[Pos96] *Robert M. Poston.* Automating Specification-Based Software Testing. *Institute of Electrical & Electronics Engineering, 1996.*

[Pos98] *Robert Poston. Making Test Cases from Use Cases Automatically. In* Proceedings of Quality Week Europe 1998*, 1998.*

[PPV00] *D.E. Perry, A.A. Porter, and L.G. Votta. Empirical studies of software engineering: a roadmap.* Proceedings of the conference on The future of Software engineering*, pages 345–355, 2000.*

[Pre01] *L. Prechelt.* Kontrollierte Experimente in der Softwaretechnik: Potenzial und Methodik. *Springer, 2001.*

[RB94] *Carol A. Reeves and David A. Bednar. Defining Quality: Alternatives and Implications.* The Academy of Management Review, Special Issue: Total Quality*, 19(3):419–445, 1994.*

[RB01] *E. Rahm and P.A. Bernstein. A survey of approaches to automatic schema matching.* The VLDB Journal The International Journal on Very Large Data Bases*, 10(4):334–350, 2001.*

[Ric03] *Filippo Ricca.* Analysis, Testing and Re-structuring of Web Applications. *PhD thesis, Universität Genua, 2003.*

[RJB04] *James Rumbaugh, Ivar Jacobson, and Grady Booch.* The Unified Modeling Language Reference Manual. *Addison Wesley Professional, 2nd edition, July 2004.*

[Roy87] *W. W. Royce. Managing the development of large software systems: concepts and techniques. In* ICSE '87: Proceedings of the 9th international conference on Software Engineering, *pages 328–338, Los Alamitos, CA, USA, 1987. IEEE Computer Society Press.*

[RT01] *Filippo Ricca and Paolo Tonella. Analysis and Testing of Web Applications. In* ICSE, *2001.*

[RvRP80] *SE Robertson, CJ van Rijsbergen, and MF Porter. Probabilistic models of indexing and searching.* Proceedings of the 3rd annual ACM conference on Research and development in Information Retrieval, *pages 35–56, 1980.*

[Rys03] *Johannes Ryser.* Szenariobasiertes Validieren und Testen von Softwaresystemen. *PhD thesis, Wirtschaftswissenschaftliche Fakultät der Universität Zürich, 2003.*

[SB88] *G. Salton and C. Buckley. Term-weighting approaches in automatic text retrieval.* Information Processing and Management: an International Journal, *24(5):513–523, 1988.*

[SCC06] *Charles Simonyi, Magnus Christerson, and Shane Clifford. Intentional software. In* OOPSLA '06: Proceedings of the 21st annual ACM SIGPLAN conference on Object-oriented programming languages, systems, and applications, *pages 451–464, New York, NY, USA, 2006. ACM Press.*

[Sco05] *Michael L. Scott.* Programming Language Pragmatics. *Morgan Kaufmann Publishers Inc., San Francisco, CA, USA, 2005.*

[SDGR03] *Ina Schieferdecker, Zhen Ru Dai, Jens Grabowski, and A. Rennoch. The UML 2.0 Testing Profile and its relation to TTCN-3. In* TestCom 2003 : testing of communicating systems, *volume 2644 of* Lecture Notes of Computer Science, *pages 79–94, 2003.*

[SEN90] *L. Schamber, M.B. Eisenberg, and M.S. Nilan. A re-examination of relevance: Towards a dynamic, situational definition.* Information Processing and Management, *26(6):755–776, 1990.*

[SFW82] *Gerard Salton, Edward A. Fox, and Harry Wu. Extended Boolean Information Retrieval. Technical report, Cornell University, Ithaca, NY, USA, 1982.*

[Sim95] *Charles Simonyi. The Death Of Computer Languages, The Birth of Intentional Programming. Technical report, Microsoft Research, 1995.*

[SK06] *Motoshi Saeki and Haruhiko Kaiya. On Relationships among Models, Meta Models and Ontologies. In* Proceedings of 6th OOPSLA Workshop on Domain-Specific Modeling (DSM'06), *number 37 in Computer Science and Information System Reports, Technical Reports. University of Jyväskylä, 2006.*

[SKS06] *Carsten Sensler, Michael Kunz, and Peter Schnell. Testautomatisierung mit modellgetriebener Testskript-Entwicklung.* Objektspektrum, *3:74–84, 2006.*

[Spi01] *Diomidis Spinellis. Notable design patterns for domain-specific languages.* The Journal of Szstems and Software, *56:91–99, 2001.*

[Spi03] *Diomidis Spinellis. On the Declarative Specification of Models.* IEEE Computer, *20(2):93–95, 2003.*

[SS08] *D. Seifert and J. Souquieres. Using UML Protocol State Machines in Conformance Testing of Components. Technical Report 00274383, Institut National Polytechnique de Lorraine Institut National Polytechnique de Lorraine, 2008.*

[Sta73] *Herbert Stachowiak.* Allgemeine Modelltheorie. *Springer, Wien, 1973.*

[Ste03] *Maximilian Stempfhuber.* Objectorientierte Dynamische Benutzungsoberflächen - Odin:Behandlung semantischer und struktureller Heterogenität in Informationssystemen mit den Mitteln der Softwareergonomie. *PhD thesis, Universität Koblenz-Landau, 2003.*

[SV06] *Thomas Stahl and Markus Völter.* Model-Driven Software Development. *Wiley & Sons, 2006.*

[TBPY01] *Wei-Tek Tsai, Xiaoying Bai, Raymond A. Paul, and Lian Yu. Scenario-Based Functional Regression Testing. In* Computer Software and Applications Conference (COMPSAC), *pages 496–501, 2001.*

[Tho04] *Dave Thomas. MDA: Revenge of the Modelers or UML Utopia?* IEEE Softwware, *21(3):15–17, 2004.*

[Thu07] *M. Thurmeier. Erwartungen an wissenschaftliche Fachportale : Ergebnisse einer qualitativen Befragung von Wissenschaftlern. Technical report, in to mind: Institut für Marketingforschung, 2007.*

[Tic00] *Walter F. Tichy. Hints for Reviewing Empirical Work in Software Engineering.* Empirical Softw. Engg., *5(4):309–312, 2000.*

[TK04] *Juha-Pekka Tolvanen and Steven Kelly. Domänenspezifische Modellierung.* OBJEKTspektrum, *04:30–35, 2004.*

[TLPH95] *W.F. Tichy, P. Lukowicz, L. Prechelt, and E.A. Heinz. Experimental evaluation in computer science: A quantitative study.* The Journal of Systems & Software, *28(1):9–18, 1995.*

[TR03] *Juha-Pekka Tolvanen and Matti Rossi. MetaEdit+: defining and using domain-specific modeling languages and code generators. In* OOPSLA '03: Companion of the 18th annual ACM SIGPLAN conference on Object-oriented programming, systems, languages, and applications, *pages 92–93, New York, NY, USA, 2003. ACM.*

[TTK04] *Antti Tevanlinna, Juha Taina, and Raine Kauppinen. Product Family Testing: a Survey.* SIGSOFT Softw. Eng. Notes, *29(2):12–12, 2004.*

[Ufl05] *Matthias Uflacker. Entwicklung eines Editors für die modellgetriebene Konstruktion komponentenbasierter Software-. Master's thesis, Universität Oldenburg, Department für Informatik, Januar 2005.*

[Uhl06] *Axel Uhl. Model-Driven Architecture. In Ralf Reussner and Wilhelm Hasselbring, editors,* Handbuch der Software-Architektur, *pages 103–128. dpunkt, 2006.*

[UPL05] *M. Utting, A. Pretschner, and B. Legeard. A Taxonomy of Model-Based Testing. Technical report, Department of Computer Science The University of Waikato, 2005.*

[vDK02] *A. van Deursen and P. Klint. Domain-Specific Language Design Requires Feature Descriptions.* Journal of Computing and Information Technology, *10(1):1–17, 2002.*

[vDKV00] *Arie van Deursen, Paul Klint, and Joost Visser. Domain-Specific Languages: An Annotated Bibliography.* SIGPLAN Notices, *35(6):26–36, 2000.*

[VG07] *Markus Voelter and Iris Groher. Product Line Implementation using Aspect-Oriented and Model-Driven Software Development. In* 11th international Software Product Line Conference 2007, *2007.*

[vL01] *Axel van Lamsweerde. Goal-Oriented Requirements Engineering: A Guided Tour. In* RE '01: Proceedings of the 5th IEEE International Symposium on Requirements Engineering, *page 249, Washington, DC, USA, 2001. IEEE Computer Society.*

[Voe03] *Markus Voelter. A collection of Patterns for Program Generation. EuroPLoP, 2003.*

[Völ05] *Markus Völter. Kaskadierung von MDSD und Modelltransformationen.* JavaSpektrum, 05/2005, *05/2005:?, 2005.*

[Völ06a] *Marcus Völter. Im Fokus: Sprachen, Modelle und Fabriken in der Softwareentwicklung.* IX Magazin für Professionelle Informationstechnik, *10:123–127, 2006.*

[Völ06b] *Markus Völter. Best Practices for Model-to-Text Transformations. In* Eclipse Summit 2006 Workshop: Modeling Symposium, *2006.*

[Völ06c] *Markus Völter. oAW xText - A framework for textual DSLs. In* Eclipse Summit 2006 Workshop: Modeling Symposium, *2006.*

[Völ07] *Markus Völter. Domain Specific Languages : Implementation Technologies, 2007.*

[Voo02] *Ellen M. Voorhees. The Philosophy of Information Retrieval Evaluation. In* CLEF '01: Revised Papers from the Second Workshop of the Cross-Language Evaluation Forum on Evaluation of Cross-Language Information Retrieval Systems, *pages 355–370, London, UK, 2002. Springer-Verlag.*

[Wei05] *S.M. Weiss.* Text Mining: Predictive Methods for Analyzing Unstructured Information. *Springer, 2005.*

[WL99] *David M. Weiss and Chi Tau Robert Lai.* Software product-line engineering: a family-based software development process. *Addison-Wesley Longman Publishing Co., Inc., Boston, MA, USA, 1999.*

[WRH+00] *Claes Wohlin, Per Runeson, Martin Hoest, Magnus Ohlsson, Björn Regnell, and Anders Wesslen.* Experimentation in Software Engineering : An Introduction. *Kluwer Academic Publishers, 2000.*

[WS06] *Dennis Wagelaar and Ragnhild Van Der Straeten. A Comparison of Configuration Techniques for Refactoring and Refinement Transformations. In Arend Rensink and Jos Warmer, editors,* Model Driven Architecture - Foundations and Applications Second European Conference, ECMDA-FA 2006 Bilbao, Spain, July 10-13, 2006, *number 4066 in Lecture Notes in Computer Science. Springer, 2006.*

[ZC04] *M.L. Zeng and L.M. Chan. Trends and issues in establishing interoperability among knowledge organization systems.* Journal of the American Society for Information Science and Technology, *55(5):377–395, 2004.*

[Zub98] *D. Zubrow. Measurement with a Focus: Goal-Driven Software Measurement.* CROSSTALK: The journal of Defense Software Engineering, *1998.*

[ZW98] *MV Zelkowitz and DR Wallace. Experimental models for validating technology.* Computer, *31(5):23–31, 1998.*

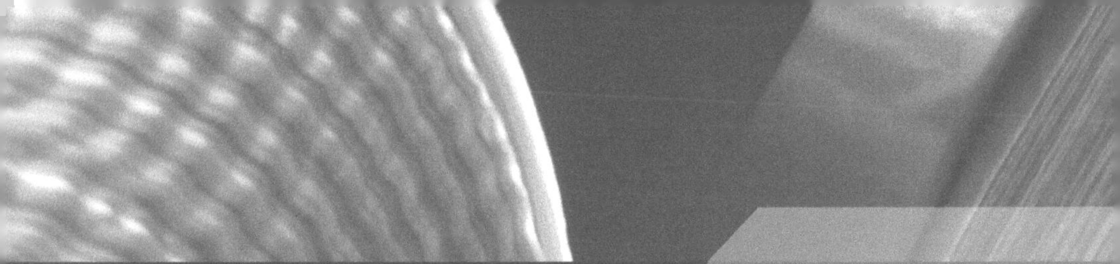

MIX
Papier aus verantwortungsvollen Quellen
Paper from responsible sources
FSC® C105338

FSC
www.fsc.org

If you have any concerns about our products,
you can contact us on
ProductSafety@springernature.com

In case Publisher is established outside the EU,
the EU authorized representative is:
Springer Nature Customer Service Center GmbH
Europaplatz 3, 69115 Heidelberg, Germany

Printed by Libri Plureos GmbH
in Hamburg, Germany